AF334960

INTERNATIONAL SERIES OF MONOGRAPHS IN

NATURAL PHILOSOPHY

GENERAL EDITOR: D. TER HAAR

VOLUME 12

**CURRENT ALGEBRAS
AND THEIR APPLICATIONS**

CURRENT ALGEBRAS AND THEIR APPLICATIONS

BY

B. RENNER

*Department of Applied Mathematics
and Theoretical Physics,
University of Cambridge, England*

PERGAMON PRESS

OXFORD · LONDON · EDINBURGH · NEW YORK
TORONTO · SYDNEY · PARIS · BRAUNSCHWEIG

Pergamon Press Ltd., Headington Hill Hall, Oxford
4 & 5 Fitzroy Square, London W.1
Pergamon Press (Scotland) Ltd., 2 & 3 Teviot Place, Edinburgh 1
Pergamon Press Inc., 44–01 21st Street, Long Island City, New York 11101
Pergamon of Canada, Ltd., 6 Adelaide Street East, Toronto, Ontario
Pergamon Press (Aust.) Pty. Ltd., Rushcutters Bay, Sydney, N.S.W.
Pergamon Press S.A.R.L., 24 rue des Écoles, Paris 5^e
Vieweg & Sohn GmbH, Burgplatz 1, Braunschweig

Library of Congress Catalog Card No. 67–24234

FILMSET BY THE EUROPEAN PRINTING CORPORATION LIMITED
DUBLIN IRELAND
PRINTED IN GREAT BRITAIN BY J. W. ARROWSMITH LTD., BRISTOL

08 003372 5

TO MY PARENTS

Contents

Contents

viii

Foreword

MORE than five years ago, Murray Gell-Mann first focussed attention on the deep significance that the algebra obeyed by the two octets (vector and axial vector) of weak interaction currents might be expected to have for elementary particle physics, a remark in line with his long standing conviction that the key to the strong interactions would be found in the study of the weak interactions. He conjectured the form which would be taken by the commutation relations between these currents, by which this $SU_3 \times SU_3$ algebra is characterized. His conjecture has at least three aspects. In the limit of exact SU_3 symmetry, the commutators of the vector currents are prescribed by the properties of the SU_3 group, since the vector currents are in one-to-one correspondence with the generators of the SU_3 group. Gell-Mann emphasized that these current commutators could still retain the same form in the presence of SU_3 symmetry-breaking interactions, even though not all of these vector currents are still conserved then, and he conjectured that this would be the case for the physical world. The second part of his conjecture, that the axial-vector currents form an SU_3 octet, is now well supported by much experimental data concerning the leptonic decay processes for mesons and baryons. The commutators between the vector and axial-vector currents are then prescribed by SU_3 symmetry. The third part of the conjecture concerns the commutation relations satisfied by the axial-vector currents and relates these commutators with the vector currents. These conjectured commutation relations are known to follow as a consequence from a number of Lagrangian theories of special form, for example from the simplest models possible for the interactions underlying the elementary particles, those based on a triplet of quark fields, as well as from many other models. The exploration of the consequences of this Current Algebra conjecture has been one of the major steps taken in the theory of the hadronic particles in recent years.

It was only two years ago that Adler and Weisberger independently realized how to use this Current Algebra to derive a sum rule for the axial vector coupling constant appropriate to beta decay for the nucleon. The value 1.18 known for the ratio $(-G_A/G_V)$ had been an outstanding puzzle in the theory of weak interactions for many years, surprisingly and tantalizingly close to the value unity known to hold for the corresponding weak interaction for leptonic particles. The Adler–Weisberger sum rule expressed this ratio in terms of pion-nucleon scattering cross-sections, albeit for pions of zero mass, since their derivation involved appeal to the hypothesis of PCAC (partially-conserved axial-vector current), which (as shown by Nambu) can be exact only in the limit of zero pion mass. Hence, the evaluation of this sum rule in terms of the physical data depended on an additional assumption of smoothness for the pion-nucleon scattering matrix-elements as function of the external pion mass (together with allowance for the variation of kinematic factors in the approach to this limit). With this reasonable assumption, there was quite remarkable agreement found between the calculated value and the observed value for G_A/G_V. This agreement was the first and dramatic demonstration of the power and deep significance of this Current Algebra, which then stimulated a great range of applications of various Current Algebra methods to problems arising from the weak interaction processes observed physically, from electromagnetic processes involving hadronic states, and from many aspects of strong interaction processes.

This wide range of applications for Current Algebra methods may be roughly separated into several categories:

(i) *Sum rules*, for which the Adler–Weisberger relation is the prototype. The most satisfactory derivation for such sum rules has been through the dispersion-theoretic methods developed by Fubini and his collaborators. The practical application of these sum rules generally involves the smoothness hypothesis mentioned above, although there are some sum rules for electromagnetic or weak interaction processes which do not depend on this hypothesis.

(ii) *Threshold theorems* for "soft pions". These are exact statements concerning the values of the matrix-element for some physical process in particular configurations where one external pion has zero four-momentum (and hence zero mass). The Kroll–Ruderman result long known (from the requirements of gauge invariance rather than of Current Algebra) for the pion photoproduction amplitude provides a typical example for such a threshold theorem. Many such results have now become established from the Current Algebra hypothesis, for many kinds of process. One remarkably successful series of applications has provided an excellent account of the K-meson decay processes, both leptonic and non-leptonic, in good general agreement with experiment. Of course, the physical applications of these theorems always involve an element of uncertainty, in that they make a precise statement only about extrapolations to unphysical configurations. However, to date, the most naive extrapolation assumptions have repeatedly proved to lead to rather reasonable results. In many physical situations now being investigated, an explicit theorem can be derived only for an approximate model for the process; in such cases, comparison of the experimental data with the prediction of this theorem may be used to provide a quantitative test for this model.

(iii) *Effects of symmetry-breaking interactions,* as in the discussion of the Gell-Mann — Okubo mass formula for hadrons by Current Algebra methods.

The literature in this field of research has developed very rapidly in the past two years, in all these directions and others besides. As a result, there has been a clear and urgent need for a book which could provide the young research student, or indeed anyone not experienced with the field-theoretic formalism of elementary particle physics, with an introduction and a guide to the methods and achievements of this Current Algebra.

As a research student at Cambridge University, Dr. Bruno Renner became known for his clear-cut and pertinent contributions to the development of the methods and applications of the Current Algebra. He was invited to give a course of lectures on the methods and applications of Current Algebra at the Rutherford High Energy Laboratory as part of our 1966 summer program in theoretical physics. In view of the widespread interest in these methods and their applications, and in their significance for the future development of elementary particle physics, it was a natural suggestion that these carefully prepared lectures should be expanded and made available for a wider audience, in the form of this book.

The book serves at least several purposes. It provides an introduction to the underlying philosophy, and to the technical methods, associated with the use of the Current Algebra for the investigation of questions in elementary particle physics. The introductory steps in the various approaches to these applications have been worked through in detail, and a variety of representative situations have been explored explicitly. The book also provides an introductory guide to the topics which have already been discussed in the literature, although it makes no attempt to review the literature completely or in detail, apart from an appendix listing about 500 papers recently published or at present in circulation. At the end of each chapter, Dr. Renner has given a brief survey of the related or parallel applications to be found in the literature, indicating their connection with those applications he has discussed in detail. As a result, the student should be able to gain a rather clear impression of the scope and possibilities for application of current algebra methods, at least as they are known at the present time, and the experimenter can find some indications of what kind of result he may expect, and where he may begin to seek it, for the processes of particular interest to him.

This book comes at a most opportune time, as the range of successful applications for the Current Algebra continues to grow, and as basic questions concerning the scope of its implications begin to be investigated. I believe that many young people, and other beginners in this field, will find that this book will provide them with a convenient and exceedingly helpful introduction into active research depending on Current Algebra methods, and that, in this way, this book will significantly influence the future work in this field.

R.H. DALITZ
Rutherford High Energy Laboratory
Chilton

1 May 1967
Department of Theoretical Physics
Oxford

Preface

OVER the last two years current algebras have been one of the most interesting fields of research in theoretical high-energy physics. The basic ideas had been proposed by Gell-Mann already in 1961, but technical difficulties delayed the applications until 1965, when Fubini and Furlan proposed workable methods and Adler and Weisberger derived their celebrated sum rule. Since this time about 500 contributions to the theory of current algebras have been published, covering a wide range of phenomena in strong, electromagnetic and weak interactions. The first generation of reviews came out in summer 1966, mainly in the form of lecture notes. Though several aspects of the theory are still under discussion, the need for a more comprehensive review became apparent.

This book is based on lectures which I gave at the Rutherford High Energy Laboratories in July 1966; the notes have been circulated as Rutherford Laboratory Report RHEL/R 126. In writing this review I found that the majority of the sections in the lecture notes could be further developed by adding new aspects of the problems and new results; some chapters have been entirely rewritten or added. To make the presentation more understandable I used the help of first-year research students whom I wish to thank on this occasion.

The purpose of a preface is also to state clearly the limitations of the book. Though this review is larger than any other one known to me, it is not complete; not every useful detail in the literature has been discussed. Had I attempted such a completeness at the time of writing, it would have been lost by the time of publication. This review is not final, because there is still discussion about several points in the theory of current algebras and we are far from having experimental tests for all its aspects. This review is not rigorous in its mathematical methods; we remain on the level of the original papers because the development of a rigorous formulation of statements in quantum field theories is a difficult and slow process. Finally, this review deals with currents algebras and not with the whole of high energy physics; we use results of dispersion theory, Lie algebras, symmetry models, Regge poles, etc., without justifying them, but we try to state these results clearly and to give references to reviews.

Over a year's studies on current algebras are reviewed in this book. At this occasion I wish to express my sincere gratitude to all scientists who helped me to develop my concepts by stimulating lectures, seminars and discussions. In particular I wish to thank my supervisor, Dr. J. C. Polkinghorne, for encouraging and stimulating guidance of my work, further I wish to thank Prof. M. Gell-Mann, Prof. S. Fubini, Dr. G. Furlan, Prof. D. Amati, Dr. J. S. Bell, Prof. J. Prentki, Prof. L. A. Radicati and many others. I wish to thank my colleagues at Cambridge where the first precursor of this review was developed in a students' seminar, at the Rutherford Laboratory where the lectures were given, and at CERN where the manuscript was completed. I have enjoyed the help of Dr. D. Sutherland and of P. Osborne, A. T. Sudbery and P. Weisz, who deserve my particular gratitude. But none of these scientists should be associated with any of the weaknesses of this presentation and certainly I cannot use their authority to support disputable statements.

Finally I wish to thank Prof. R. H. Dalitz, without whose encouraging suggestions this book would never have been written, and Pergamon Press for their very friendly and efficient co-operation.

Some technical remarks now on the order of presentation. The main text is followed by two appendixes: (I) on LSZ-reduction formulae and (II) on the notation used in this book. The list of references is divided into a current algebra bibliography, referred to by simple numbers in parentheses, a supplement to the bibliography added in proof and referred to only in footnotes, and some selected general references, referred to by a number preceded by "S". Equations are numbered consecutively in each chapter; (4.14), for instance, refers to the fourteenth equation of chapter 4.

CHAPTER 1

Preliminary Concepts

1.1. The Interactions of Subnuclear Particles

With our present knowledge of high-energy physics we can describe the reactions of subnuclear particles by four or five different interactions: the strong interactions of baryons and mesons,† the electromagnetic interactions mediated by photons, the weak interactions observed mainly in slow decay processes and neutrino reactions, and gravitation. The recently discovered CP-violation may be an effect of a possible fifth interaction. There may also be others, or the phenomenological distinction of different interactions may not be fundamental at all and may at some stage no longer be useful.

The present concept of different interactions is based on two classes of phenomena:

(a) In cross-sections, decay rates and mass-differences we find characteristic orders of magnitude for the effects of each inter-action. Universal coupling constants have been identified for the electromagnetic and weak interactions: the electric charge e and the Fermi coupling constant G. For strong interactions the pion-nucleon coupling constant g represents the order of magnitude, but no universal role is ascribed to it. Gravitation coupling is universal at the macroscopic level, but it is too weak to be tested with subnuclear particles. We will not consider it further. Similarly we neglect CP-violation because although we do not know enough about it, there are reasons to believe that its effects are small.

Baryons and mesons take part in all interactions; of the leptons, electrons and muons participate in electromagnetic and weak interactions, neutrinos only in weak interactions.

---

†In the following we will often use the term hadrons to denote baryons and mesons.

(b) The interactions have different symmetries and conservation laws. Relativistic invariance in respected by all interactions, and the product operation CPT (charge conjugation, space reflexion, time reversal) is taken to be a universal symmetry. Absolute conservation laws have been found for the baryon number B, lepton number L† and electric charge Q. Ascribing the recently discovered very weak violation of CP- (and consequently T-) invariance in the decay process $K° \rightarrow \pi^+\pi^-$ to a new interaction, then the following quantum numbers are conserved:

strong interactions: B, L, Q, C, P, T, S (strangeness), I (isospin);
electromagnetic interactions: B, L, Q, C, P, T, S;
weak interactions: $B, L, Q, (CP), T$.

Only by the existence of leptons and by the violation of symmetries of the stronger interactions have the weaker ones been separated out.

The description of these interactions has always been based directly or indirectly on some form of relativistic causal quantum mechanics. A sequence of approximations can be designed: first we consider only strong interactions and consequently disregard photons and leptons. This has the advantage that there are only massive particles left to deal with. As a consequence of Heisenberg's energy–time uncertainty principle any forces produced by the exchange of only massive quanta must have a finite range. Thus one can introduce asymptotically free states of stable particles and describe scattering processes as transitions among them by a unitary S-matrix. The asymptotic states can contain only a finite number of particles at any given energy, so the transition amplitudes always depend only on a finite number of variables. It has been a very fruitful assumption to regard them as analytic functions, giving various kinds of dispersion relations as integrals over Cauchy contours. The analyticity assumption is now regarded as an expression of causality principles. In the use of dispersion relations some additional assumptions have always to be made, such as crossing principles to interpret singularities outside the kinematically allowed domain of variables, assumptions about asymptotic properties in complex planes and for practical calculations dominance of

---

†If our ideas about weak interactions are correct, we have a separate conservation of the electron lepton number L_e and the muon lepton number L_μ.

certain known singularity-contributions over unknown ones. We cannot go into details here, but we refer to some selected literature both on the research to justify the analyticity assumptions from causal relativistic quantum field theory, and on the analysis of the singularities of the S-matrix and their use in practical calculations.[S 1–S 6]

Dispersion relations do not allow us to calculate scattering amplitudes from an input of a few fundamental constants or physical laws, they are an expression of analytic self-consistency and can only connect observed data of different reactions with each other; these connections are used for tests and predictions with good success. Nevertheless, there remains the question of how much freedom there is in a maximally analytic S-matrix, to what extent analytic self-consistency determines the S-matrix, or how much additional information is necessary as an input to construct from it the S-matrix of strong interactions only by analyticity principles. This question, proposed not on the level of practical calculations but as a matter of principle, is not likely to be answered in the near future, mainly because of the mathematical difficulties of inelastic reactions. The bootstrap hypothesis[S 5] suggests that in principle the full S-matrix can be determined from postulates of self-consistency and least singularity. Alternatively if one believes quark models[S 7] one may think that very essential dynamical details are necessary as input.

The symmetries and approximate symmetries of strong interactions will play an important role in this discussion. Different points of view are presented about their origin. Current algebras may be seen as a theory connecting these seemingly diverse developments.

The next step of approximation would be to take into account electro-magnetic and weak interactions in first order of their coupling constants e and G. This step introduces the leptons, but in this approximation their Hilbert space of states is a space of free particles, describable by local fields, and completely separated from the hadron space. Effective Hamiltonians are used to describe electromagnetic and weak reactions to lowest order

$$\mathcal{H}^{\mathrm{el}} = e j_\mu^{\mathrm{el}} A^\mu, \qquad (1.1)$$

$$\mathcal{H}^W = \frac{1}{\sqrt{2}} G \frac{1}{2} [J_\mu^W (J^{\mu W})^+ + (J^{\mu W})^+ J_\mu^W]. \qquad (1.2)$$

3

The Hamiltonian of electromagnetic interactions (1.1) is a product of the photon field operator A^μ with the electromagnetic current operator j_μ^{el}. A classical analogue, the interaction energy of charges and radiation, has facilitated the proposal of this form. The matrix elements of j_μ^{el} always have the general form of a conserved vector current; for example with nucleons of momenta p^i and p^f:†

$$\langle N(p^f)|j_\mu^{el}(x)|N(p^i)\rangle = [\exp i(p^f - p^i)x][\bar{u}(p^f)(\gamma_\mu F_1^{el}(q^2) + \sigma_{\mu\nu}(p^i - p^f)_\nu F_2^{el}(q^2))u(p^i)]$$

with $\qquad \sigma_{\mu\nu} = \dfrac{1}{2}(\gamma_\mu\gamma_\nu - \gamma_\nu\gamma_\mu)$ and $q^2 = (p^f - p^i)^2 \qquad$ (1.3)

$F_1^{el}(q^2)$ and $F_2^{el}(q^2)$ are the Pauli–Dirac form factors. Classically they can be considered as Fourier transforms of extended charge and magnetic moment distributions; $F_1^{el}(0)$ is the total charge, $F_2^{el}(0)$ is the anomalous magnetic moment μ^A. The existence of non-trivial form factors can be explained physically by virtual‡ strong interactions of the nucleon with pions and other states giving rise to a cloud structure.

Leptons have a similar cloud structure, but only due to higher order electromagnetic effects. In first order of e they appear structureless, therefore we take constant form factors for (point-like) leptons:

$$\langle l(p^f)|j_\mu^{el}(x)|l(p^i)\rangle = [\exp i(p^f - p^i)x] \cdot Q_l \cdot \bar{u}(p^f)\gamma_\mu u(p^i) \quad (1.4)$$

with the lepton charge $Q_l = \pm 1, 0$.

This allows us to use a description of the lepton electromagnetic current in terms of local fields:

$$j_\mu^{el,l}(x) = [\psi^{(e)}(x)]^+\gamma_0\gamma_\mu\psi^{(e)}(x) + [\psi^{(\mu)}(x)]^+\gamma_0\gamma_\mu\psi^{(\mu)}(x) \quad (1.5),$$

where $\psi^{(e)}(x)$ is introduced as annihilating a positron and creating an electron, $\psi^{(\mu)}(x)$ is the analogous muon field.

---

†Unless stated differently, we will always use invariant normalization:
$$\langle p^f|p^i\rangle = (2\pi)^3 2E_p\delta^3(p^f - p^i).$$
‡The separation of interaction partners from the nucleon should be understood through uncertainty relations in time and energy.

For weak interactions the Hamiltonian (1.2) had to be guessed and to be compared with experiment; in fact (1.2) is still not yet established for non-leptonic weak reactions. $\mathcal{H}^W$ is a symmetrized product of the weak current J_μ^W with its hermitean adjoint. It contains a lepton- and a hadron-part (like the electromagnetic current), but each of these has a vector and an axial vector component. Consequently $\mathcal{H}^W$ is not invariant under a space reflection, but in the light of our present knowledge it may be taken to be invariant under the combined operation of space reflexion and charge conjugation (CP). The symmetrization of the product (1.2) expresses this invariance.[S 8]

The lepton weak current $J_\mu^{W,l}$ has a simple representation with electron, muon and neutrino fields:

$$J_\mu^{W,l}(x) = [\psi^{(e)}(x)]^+\gamma_0\gamma_\mu(1 - \gamma^5)\psi^{(\nu e)}(x)$$

$$+ [\psi^{(\mu)}(x)]^+\gamma_0\gamma_\mu(1 - \gamma^5)\psi^{(\nu\mu)}(x). \qquad (1.6)$$

$\psi^{(\nu e)}(x)$ is introduced as annihilating an electron-antineutrino and creating an electron-neutrino, $\psi^{(\nu\mu)}(x)$ is introduced similarly.

For the hadron part we have to use again the most general possible vector and axial vector currents with form factors representing the cloud effects of strong interactions. For instance in β-decay: $N \rightarrow P + e^- + \nu_e$ we would have the matrix element

$$\langle P(p^f)|J_\mu^W(0)|N(p^i)\rangle = \bar{u}_P(p^f)(F_{1,PN}^W(q^2)\gamma_\mu$$

$$+ F_{2,PN}^W(q^2)\sigma_{\mu\nu}(p^i - p^f)_\nu + F_{3,PN}^W(q^2)(p^i - p^f)_\mu + G_{1,PN}^W(q^2)\gamma^5\gamma_\mu$$

$$+ G_{2,PN}^W(q^2)\gamma^5\sigma_{\mu\nu}(p^i - p^f)_\nu + G_{3,PN}^W(q^2)\gamma^5(p^i - p^f)_\mu)u_N(p^i). \qquad (1.7)$$

The form factors represent again the cloud effects of strong interactions. We will see that the weak vector current form factors will be related to the electromagnetic form factors. This relation can be described very elegantly and will form the starting point for current algebras; but with our present knowledge we cannot give a deeper reason why such a connection should exist between as yet unrelated interactions.

Eventually the sequence of approximations should be continued to include higher order electromagnetic and weak effects. For electromagnetism this problem has been largely solved by renormalized quantum electrodynamics, for weak interactions it is as yet unsolved. In this book we will remain on an intermediate level: we will consider strong interactions exactly and will treat them by S- matrix methods in a Heisenberg picture; electromagnetic and weak interactions will be considered only in lowest order as causing transitions between the eigenstates of strong interactions, described by the Hamiltonians (1.1) and (1.2). The Hilbert spaces of hadrons and leptons remain completely separated, and only hadrons need be considered in summations over intermediate states.

1:2. The Isospin Symmetry of Strong Interactions

The only exact internal symmetry group of strong interactions is the isospin group. Heisenberg has proposed the following analogy: In the absence of a magnetic field no physical result (of non-relativistic quantum mechanics) changes when the spin of an electron is rotated, i.e. when its spin wave function χ is transformed into $M\chi$, where M is a two-dimensional unitary and unimodular† matrix; the same kind of transformation between proton and neutron states in an isospinor $-\psi = \begin{pmatrix} P \\ N \end{pmatrix}$ into $M\psi$ — should change no physical result when electromagnetic and weak interactions are neglected and only strong interactions considered. The small nucleon mass splitting is taken as an electromagnetic effect. The group of transformation matrices M is known as SU_2; it is locally isomorphic to the rotation group and has irreducible unitary representations in spaces of any dimension. In the neighbourhood of the identity an SU_2 transformation U can be represented by the Lie generators I^i:

$$U = \exp\left(i\alpha_i I^i\right) \approx 1 + i\alpha_i I^i \quad (i = 1,2,3). \tag{1.8}$$

When U is interpreted as a rotation in a three-dimensional space the operators I^i are the familiar angular momentum operators. Here we use U as a transformation between different particles and call I^i the isospin operators, but this does not affect their group properties. As the angular momentum operators, the isospin operators have the commutators:

† Matrices with determinant one are called unimodular.

6

$$[I^i, I^j] = ie_{ijk}I^k. \tag{1.9}$$

These commutators follow from the group structure, as the commutators of angular momentum operators follow from the fact that the product of two rotations depends on their order. They are sufficient to characterize the group locally because the transformations U can be reconstructed from I^i by a generalized integration. As a magnetic field distinguishes a direction in ordinary space, so do electromagnetic interactions distinguish a direction in isospace by $Q = I^3 + Y/2$ (hypercharge $Y = S + B$). Proton and neutron can now be considered as an isospin doublet of a nucleon state. All hadrons can be grouped into isospin multiplets of approximately equal mass, and, moreover, their strong interaction dynamics do indeed turn out to be invariant under isospin transformations.

Mathematically we describe this invariance as follows: In the Hilbert space of hadron states there exists an SU_2-group of unitary transformations U, which commute with the generators of the Poincaré group. The operators U are time-independent, because:

$$i\frac{dU}{dt} = [P_0, U] = 0 \Rightarrow i\frac{dI^i}{dt} = [P_0, I^i] = 0. \tag{1.10}$$

We find the following consequences:

(a) States of given energy and momentum p_μ must always occur in complete SU_2 representations because $[P_\mu, U] = 0$. For single-particle states this implies mass-degenerate SU_2 multiplets. Considering rotations R in the rest frame: $[R, U] = 0$, we conclude that states in a multiplet must have the same spin.

(b) The S-matrix is unchanged by simultaneous isospin transformations of initial and final states:

$$S_{\beta\alpha} = \langle \beta^{t=+\infty} | \alpha^{t=-\infty} \rangle = \langle \beta^{t=+\infty} | U^{-1} U | \alpha^{t=-\infty} \rangle$$
$$= \langle (U\beta)^{t=+\infty} | (U\alpha)^{t=-\infty} \rangle. \tag{1.11}$$

This implies selection rules and relations between transition amplitudes; for instance, in the pion–nucleon system there are only two independent transition amplitudes corresponding to total isospin $\frac{3}{2}$ and $\frac{1}{2}$, describing eight possible reactions.

(c) One often finds that physically interesting operators have a simple transformation behaviour under a symmetry group: they form a number of operator multiplets. An operator multiplet is a set of operators which are transformed into each other with the same matrices (α_{ik}) as the members of a state multiplet:

$$|a_i\rangle \Rightarrow U|a_i\rangle = \sum_k \alpha_{ik}|a_k\rangle \quad \text{and} \quad \hat{O}_i \Rightarrow U^{-1}\hat{O}_i U = \sum_k \alpha_{ik}\hat{O}_k.$$

To characterize the transformation properties of these so-called tensor operators it is not necessary to work with finite transformations U. It is sufficient to know the commutators of the tensors with the Lie generators. For instance, an isovector-triplet is characterized by:

$$[I^i,\hat{O}^j] = ie_{ijk}\hat{O}^k \tag{1.12}$$

The electromagnetic current is taken to be the sum of an isovector and an isoscalar current, as suggested by the relation $Q = I^3 + Y/2$. The isospin properties of the hadron weak current $J_\mu^{W,h}$ have to be found from experiment, in particular from hadron leptonic decays. The present experimental evidence is consistent with regarding the strangeness conserving part J_μ^{I} as a member of an isospin triplet, the strangeness changing part J_μ^{II} as a member of an isospin doublet. We split:

$$J_\mu^{W,h} = J_\mu^{I} + J_\mu^{II} \tag{1.13}$$

with quantum numbers

$$Q = 1, S = 0, I^3 = 1, I = 1; \quad Q = 1, S = 1, I^3 = \tfrac{1}{2}, I = \tfrac{1}{2}$$

inducing hadron transitions with:

$$\Delta Q = 1, \Delta S = 0, \Delta I^3 = 1, \Delta I = 0, \pm 1;$$

$$\Delta Q = 1, \Delta S = 1, \Delta I^3 = \tfrac{1}{2}, \Delta I = \pm\tfrac{1}{2}.$$

The isospin properties of a class of operators can be tested with the help of the Wigner–Eckart theorem. Having assigned quantum numbers $(I,I^3) = (J,M)$ to the members of an operator

8

multiplet $\hat{O}(J,M)$, their matrix elements between states $\langle\alpha(j',m')|$ and $|\beta(j,m)\rangle$ of arbitrary isospin multiplets can be computed in terms of the Clebsch–Gordan coefficients of the rotation group (addition of angular momenta) up to a common scaling factor $\langle j'\|J\|j\rangle$, the so-called reduced matrix element, which is independent of m,m' and M.†

$$\langle\alpha(j',m')|\hat{O}(J,M)|\beta(j,m)\rangle = \langle j'\|\hat{J}\|j\rangle C(j'm'|J,M;j,m).$$

$$(1.14)$$

An example at present under experimental test are the decays $(K^+ \to \pi^0 + \text{leptons})$ and $(K^0 \to \pi^- + \text{leptons})$, where (1.13) and (1.14) would require

$$\langle\pi^-|J_\mu^{\mathrm{II}}|K^0\rangle = \sqrt{2}\langle\pi^0|J_\mu^{\mathrm{II}}|K^+\rangle \qquad (1.15)$$

because J_μ^{II} is proposed to be an isospinor. Equation (1.15) implies not only a relation between the decay rates, but also a proportionality of the form factors.

It is a matter of theoretical speculation and experimental tests which operators can be grouped into the same isospin multiplet. The vector part j_μ^1 of the strangeness conserving hadron weak current J_μ^1 (eq. (1.13)) is the ($I^3 = +1$)-member of an isotriplet of currents, its hermitean adjoint is the ($I^3 = -1$)-member, the neutral member is so far unidentified. The isovector part of the electromagneitc current j_μ^v is the neutral member of some triplet of currents, but so far we have not identified its charged members. The important suggestion of Gerstein and Zeldovitch[S 10] and Feynman and Gell-Mann[S 11] is to assign these three currents to the same triplet and to make them isospin partners of each other.

$$j_\mu^1 = (j_\mu^1 + i j_\mu^2); \quad (j_\mu^1)^+ = (j_\mu^1 - i j_\mu^2); \quad j_\mu^v = j_\mu^3$$

$$\text{with} \quad [I^i, j_\mu^j(x)] = i e_{ijk} j_\mu^k(x) \qquad (1.16)$$

†In higher groups like SU_3 the Wigner–Eckart theorem is modified and sometimes more than one undetermined scaling factor exists. An important example is the coupling of three octets which contains two free constants F and D:

$$\langle\alpha(8,i)|\hat{O}(8,j)|\beta(8,k)\rangle = i f^{ijk} F + d^{ijk} D$$

in Gell-Mann's notation.[S 9]

9

This assignment has two main consequences:

(1) Like the isovector part of the electromagnetic current j^v_μ, the part j^{I}_μ of the hadron weak current is conserved: $\partial^\mu j^{\mathrm{I}}_\mu = 0$, and the space-integral over its time component is a time-independent operator: $I^i = \int (dx)^3 j^i_0(x)$. This conservation property allows us to fix the scale of j^{I}_μ uniquely, because $\int (dx)^3 j^{\mathrm{I}}_0(x)$ satisfies the non-linear relations (1.9), and we can understand, at least in perturbation theoretic models, why the universal weak coupling constant G in nuclear β-decay is not altered by the presence of strong interactions and, like the charge e, has kept the same value as for leptons.† This is why the theory has become known as conserved vector current theory (CVC).

(2) The form factors of the weak vector current and the isovector part of the electromagnetic current should be the same. For nucleons this implies, for instance, $F^W_{3,PN}(q^2) \equiv 0$ in (1.7) as a consequence of current conservation and $F^W_{2,PN}(0) = (\mu^A_P - \mu^A_N)$, fixed by the anomalous nucleon magnetic moments. By precise measurements of the electron energy spectra in nuclear β-decays, a term $\sigma_{\mu\nu}(p^N - p^P)_\nu(\mu^A_P - \mu^A_N)$ in (1.7) has indeed been confirmed (weak magnetism).[S 13]

We summarize this section by two independent physical statements:

(a) The isospin group SU_2 is a symmetry group of strong interactions.

(b) Its Lie generators are connected with currents in electromagnetic and weak interactions, i.e. charges of observable currents satisfy the algebra: $[I^i, I^j] = ie_{ijk}I^k$ and the stronger relations: $[I^i, j^j_\mu(x)] = ie_{ijk}j^k_\mu(x)$.

1.3. The SU_3-Symmetry Model

Generalizations of the physical results in the isospin group have led to the development of higher symmetry models and current algebras. At first models were investigated with larger symmetry groups than SU_2. The idea behind these was a conjectured division of strong interactions into a dominant part conserving the higher symmetry and "semistrong" interactions breaking it. Despite some ambiguities in comparing model predictions and observed data, due to the large symmetry breaking, good approximations were obtained to many different

---

†A slight modification in scale will be made later in Cabibbo's theory.[S 12]

physical data. The most successful model of this kind has been SU_3 symmetry, suggested by Gell-Mann and Ne'eman.[S 9]

SU_3 is the group of unitary and unimodular transformations in a three-dimensional complex space; it contains isospin transformations as an SU_2 subgroup. As opposed to the isospin-case, no SU_3-triplets have been found in nature, only higher representations seem to occur; in particular an octet representation for the $(\tfrac{1}{2})^+$ baryons,† a decuplet for the $(\tfrac{3}{2})^+$ baryons (containing the celebrated Ω^-), and an octet and a singlet for each of the 0^- and 1^- mesons. To identify an SU_3 octet we note that it contains two isospin doublets, an isospin triplet and an isospin singlet. These multiplets can be distinguished by an isoscalar quantum number having the values ± 1 for the doublets and 0 for the triplet and singlet. On comparison with the hadron states one interprets this quantum number as the hypercharge $Y = S + B$. Thus in an (Y, I^3)-diagram, the $(\tfrac{1}{2})^+$ baryons and 0^- mesons fill the places of Fig. 1.

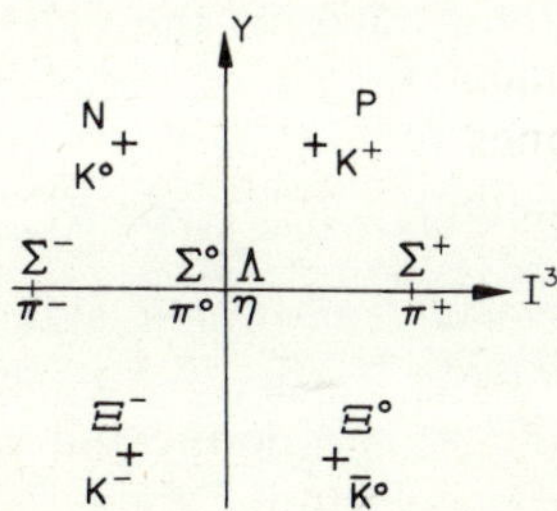

Fig. 1. Multiplets of SU_3: $(\tfrac{1}{2})^+$ and 0^-

This classification is further supported, if one speculates on the nature of the SU_3 symmetry breaking forces responsible for the appreciable mass differences between the members of SU_3 multiplets. Assuming that the Hamiltonian of strong interactions can be split as $H = H^0 + \lambda H^8$, where H^0 creates the average supermultiplet masses and is SU_3-invariant and H^8 creates the mass splittings and has the SU_3 transformation properties of an octet member, then we can apply the Wigner–Eckart theorem for expectation values $\langle B_i | H^8 | B_i \rangle$ between the octet members $|B_i\rangle$. This gives their mass splittings in first order of λ. Between

---

† $(\tfrac{1}{2})^+$ refers to spin parity J^P.

11

octet states the matrix element of an octet operator contains two free constants, so there is always one relation between the three mass differences, the so-called Gell-Mann–Okubo mass formula.

$$(m_N + m_\Xi)/2 = (3m_\Lambda + m_\Sigma)/4 \quad \text{for} \quad (\tfrac{1}{2})^+ \text{ baryons} \qquad (1.17)$$

$$m_k^2 = (3m_\eta^2 + m_\pi^2)/4 \quad \text{for} \quad 0^- \text{ mesons} \qquad (1.18)$$

In the literature linear mass formulae are normally given for baryons, quadratic ones for mesons. However, since linear and quadratic mass formulae differ only by second-order terms: $\Delta(m^2) = 2m\Delta m + (\Delta m)^2$, it is rather academic to discuss this difference here.

Another interesting part of the SU_3 symmetry model is Cabibbo's extension[S 12] of the CVC theory (Section 1.2); he proposed that the strangeness changing and the strangeness conserving parts of the hadron weak current should be members of the same octet of vector and axial vector currents: $(j_\mu^i(x) + \bar{j}_\mu^i(x))$;† for the vector currents the octet j_μ^i contains also the electromagnetic current. An angle-parameter ϑ sets the scale:

$$J_\mu^{\mathrm{I}} = \cos \vartheta (j_\mu^1 + i j_\mu^2) + \cos \vartheta (\bar{j}_\mu^1 + i \bar{j}_\mu^2), \qquad (1.19)$$

$$J_\mu^{\mathrm{II}} = \sin \vartheta (j_\mu^4 + i j_\mu^5) + \sin \vartheta (\bar{j}_\mu^4 + i \bar{j}_\mu^5), \qquad (1.20)$$

$$j_\mu^{\mathrm{el}} = j_\mu^3 + \frac{1}{\sqrt{3}} j_\mu^8. \qquad (1.21)$$

According to recent estimates $\vartheta \approx 0 \cdot 212 \pm 0 \cdot 004.$[S 14] In estimating ϑ from decay data, SU_3 violations should be taken into account. Practically this is very difficult, but general theorems[S 15] suggest that the effects of SU_3 violations are much smaller with the matrix elements of vector currents than with the matrix elements of axial vector currents. From the latter ones $\vartheta \approx 0 \cdot 268 \pm 0 \cdot 001$ has been estimated.[S 14] In current algebras we regard this discrepancy as being due to SU_3 violations and we will qualitatively connect it to other SU_3 violations (Sections 5.3 and 6.2).

The assignment (1.19) differs from (1.16) by the presence of

---

†The index i refers to the octet basis in the notation of Gell-Mann,[S 9] the bar in $\bar{j}_\mu$ denotes axial currents.

12

the factor $\cos \vartheta \approx 0.98$ in the strangeness conserving hadron weak currents, which is not present in the lepton weak currents. Very accurate comparisons between the β-decay of the neutron and the μ-meson have confirmed this difference.[S 13]

The Cabibbo angle ϑ describes the relative orientation of the weak current in SU_3 space to the direction of SU_3 symmetry breaking (Y). Its magnitude may be explained only through simultaneous considerations of strong and weak interactions in higher order; on the present level of understanding it has to be taken as an input constant of unknown origin. Its smallness has nothing to do with SU_3 violations, at least if current algebras are correct, though of course a χ^2 fit of ϑ with decay data may be quantitatively affected.

The Lie generators of the SU_3-algebra satisfy the characteristic commutation relations:

$$[Q^i,Q^j] = if^{ijk} Q^k \quad (i,j,k = 1 \ldots 8) \qquad (1.22)$$

f^{ijk} are the SU_3 structure constants.

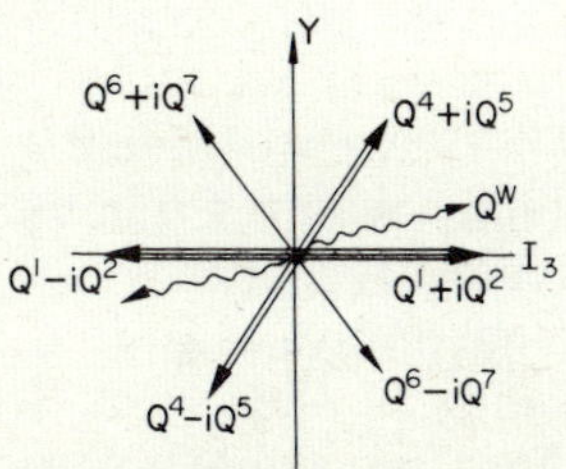

Fig. 2. The Lie generators of SU_3.

The algebra contains the isospin generators $Q^i = I^i$ for $i = 1,2,3$ and the hypercharge Y. The generators are shown in a Cartan diagram (Fig. 2), where $Q^3 = I^3$ and $Y = 2/\sqrt{3}\ Q^8$ form the Cartan centre and provide coordinate axes, characterizing the other generators by the transitions they produce. The generators can all be identified with weak or electromagnetic charges in the SU_3 symmetry model, either directly (double lines in Fig. 2) or by isospin transformations. The weak charge $Q^W = \int J_0^{W,h}(x)d^3x$ has been denoted by a wavy line; its orientation is specified by ϑ.

The difficulties and ambiguities start when quantitative predictions are to be made from SU_3. For illustration we take the weak decays $\pi^- \to \mu^- + \bar{\nu}_\mu$ and $K^- \to \mu^- + \bar{\nu}_\mu$. The hadron matrix elements are

$$\langle 0 | J_\mu^{W,h} | \pi^- \rangle = \cos \vartheta \langle 0 | (\bar{j}_\mu^1 + i\bar{j}_\mu^2) | \pi^- \rangle = i \cos \vartheta \, \frac{f_\pi^{(n)}}{(m_\pi)^n} \cdot p_\mu^\pi, \quad (1.23)$$

$$\langle 0 | J_\mu^{W,h} | K^- \rangle = \sin \vartheta \langle 0 | (\bar{j}_\mu^4 + i\bar{j}_\mu^5) | K^- \rangle = i \sin \vartheta \, \frac{f_K^{(n)}}{(m_K)^n} \cdot p_\mu^K. \quad (1.24)$$

In the SU_3 symmetric model $m_\pi = m_K$ and $f_\pi^{(n)} = f_K^{(n)}$ for every scaling. A change of the power n of the masses in the scaling factor $1/(m)^n$ produces just a trivial redefinition of the decay constants $f^{(n)}$, but it does not affect their equality. So in the SU_3 symmetric model there would be no problem in determining ϑ from the partial decay widths Γ:

$$\tan^2 \vartheta = \Gamma_{K \to \mu\nu} / \Gamma_{\pi \to \mu\nu}.$$

In reality we have to take account of the different phase spaces because $m_K \approx 3.5 m_\pi$, but we also have to decide on the scaling. Choosing $n = 0$ and $\vartheta \approx 0.212$ we find $f_K^{(0)}/f_\pi^{(0)} \approx 1.26$ or alternatively assuming $f_K^{(0)} = f_\pi^{(0)}$ we would find $\vartheta \approx 0.268$. The result that the best fit is obtained with $n = 0$ in the scaling lies outside the scope of the SU_3 symmetry model.

The ambiguities are not always as large as here and often some sort of "natural" scaling is successful. But in principle the ambiguities remain for the symmetry model. With current algebras we will partly solve them; in Section 6.2, for instance, we will show why the best fit to $f_\pi^{(n)} \approx f_K^{(n)}$ is obtained with $n = 0$.

The same scaling ambiguity turns up in comparing resonance decay rates and effective coupling constants. A new sort of ambiguity is connected with the energies at which scattering processes should be compared to each other. At 1600 MeV c.m.s. energy (πN) scattering has already exhibited three resonances while $(K\Xi)$ is still below threshold. The regions of dominance of certain dynamical processes depend in a complicated way on the masses of exchanged systems. If the latter are mixed by mass breaking among multiplet members, there may be no possible prescription as to how the energies of scattering processes should be related when comparing cross-sections. The situation with resonance production is similar. This is why SU_3 has not been very successful for scattering data; only where one sufficiently simple mechanism dominates may one expect a good fit.[S 16]

Thus it is necessary to consider carefully the violations of SU_3. Isospin symmetry is also violated by electromagnetism, but there we know the nature of the breaking forces, and further we have renormalized quantum-electrodynamics to deal with them. In the case of SU_3 we do not know the breaking forces (apart from a suggested octet behaviour), and the use of perturbation theory has so far not been very successful in strong interactions. Lagrangian models of SU_3 breaking can be constructed, but their physical meaning is problematic.

The application of SU_3 to reality contains not only practical ambiguities, but also conceptual problems. SU_3 transformations are well defined in a symmetry model, but what is an SU_3 transformation on physical states? Does it transform, for instance, single-baryon states into each other (at equal spatial momentum? —this would not be manifestly covariant; at zero relative velocity?—this would not be translationally invariant) or into states with admixtures (how to determine these admixtures?)? We see that the concept of an approximate symmetry is unclear and additional prescriptions or a different introduction of SU_3 is necessary for further advancement.

1.4. Quarks and the SU_6-Model

We start with some mathematical statements. Finite dimensional and unitary representations of compact Lie groups can always be constructed as tensors from the fundamental representations. As an example we will demonstrate this for SU_3. The fundamental representations are the triplet representations, given by unitary and unimodular matrices M and their complex conjugates M^* in a three-dimensional complex space. Its basis vectors will be denoted by q^i and q^{*i} with $i = 1,2,3$.

$$q^i \rightarrow M_{ik}q^k; \quad q^{*l} \rightarrow M^*_{lm}q^{*m}. \tag{1.25}$$

A tensor product of two basis vectors $q^i q^k$ provides a nine-dimensional representation space for SU_3

$$q^i q^k \rightarrow M_{ij}M_{kl}q^j q^l, \tag{1.26}$$

but this representation can be decomposed into a six-dimensional one with symmetrized products $s^{ik} = q^i q^k + q^k q^i$ and a three-dimensional one with antisymmetrized products $a^{ik} = q^i q^k - q^k q^i$.

$$s^{ik} \to M_{ij} M_{kl} s^{jl}; \quad a^{ik} \to M_{ij} M_{kl} a^{jl}. \tag{1.27}$$

Three basis vectors $q^i q^j q^l$ would provide a twenty-seven-dimensional representation which contains a decuplet (totally symmetric products), two octets (mixed symmetry) and a singlet (totally antisymmetric).

The SU_3 representations of physical particles, octets and decuplets, can be obtained from suitably symmetrized products of two or three triplet states q^i and q^{*j}. The question is now whether this construction has only a formal meaning or also some physical interpretation by considering the hadrons in SU_3 multiplets as bound states of more fundamental entities: quarks q^i and antiquarks q^{*j}.[S 7]

Bound states are fully understood only in non-relativistic quantum mechanics. Though one would not have guessed that anything can be learnt about hadron structure from nonrelativistic models, in fact their study turns out to be very interesting.

To build up baryons quarks are assumed to have spin $\frac{1}{2}$; baryons would consist of three quarks, mesons of a quark and an antiquark. In this connection Gürsey and Radicati[S 17] made the interesting hypothesis that to some approximation the interquark forces should not couple at all to the quark spins so that there is not only an SU_3 invariance, but also an SU_6 invariance in the static limit. (Three quark states with two spin components each give a six-dimensional state-space.) To make the concept clear consider three transformations:

(a) An SU_3 transformation leaves the spins unchanged.

(b) A rotation of the spins leaves the internal quantum numbers unchanged. Invariance against (a) and (b) would lead to an $SU_3 \times SU_2$ symmetry.

(c) We rotate the spins of the p-quarks (q^1, q^{1*}), leave the spins of the n-quarks (q^2, q^{2*}) unchanged and rotate the spins of the λ-quarks (q^3, q^{3*}) into the opposite direction. Such an operation is a unitary transformation in the six-dimensional quark space, but not a product of a universal rotation and an SU_3-transformation. Invariance against (a), (b) and (c) is implied by spin-independent forces, and this is what is meant by SU_6 invariance. (Spin-orbit forces would break the invariance against (b) and (c); spin–spin forces would break invariance against (c) but not (b).) We stress again that we are dealing here with the non-relativistic limit.

The quark–antiquark system leads to SU_6 representations of dimensions 35 and 1. The thirty-five dimensional representation contains an SU_3 octet of spin zero, an octet of spin one and a singlet of spin one. A short notation for representing an SU_6 multiplet by its $SU_3 \times SU_2$ content is

$$35 = (8,1) + (8,3) + (1,3). \tag{1.28}$$

If we carry the bound state picture further and assume an s-wave state for the quark-antiquark system then the parity of bound states would be negative, because fermions and antifermions have always negative relative parity. The spectrum of 35 exactly fits the lowest 0^- and 1^- mesons and higher mesons can be classified in multiplets of quark–antiquark states with orbital angular momentum.

The situation is more complicated with the baryons. A three-quark system gives SU_6 representations 2×70, 56 and 20. Their $SU_3 \times SU_2$ content is:

$$(70) = (10,2) + (8,4) + (8,2) + (1,2),$$
$$(56) = (10,4) + (8,2),$$
$$(20) = (8,2) + (1,4). \tag{1.29}$$

The representation 56 contains an $(\tfrac{1}{2})^+$ octet **and** a $(\tfrac{3}{2})^+$ decuplet (assigning positive parity to quarks), which well fits the lowest baryon states. Why just this multiplet is dynamically preferred is not completely understood. If Fermi–Dirac statistics are obeyed by the quarks, a completely antisymmetrical spatial wave function is required because the representation 56 is completely symmetrical in spin and SU_3 indices. This causes difficulties in understanding interquark forces, but it is not the subject of this book to review dynamical models of quark binding.[S 7]

Interesting results are obtained from SU_6 about the electromagnetic current, which is assumed to transform like an SU_6

generator and assigned to a *35*-plet of currents. This predicts the ratio of the nucleon magnetic moments very well, but a similar assumption on the weak axial currents gives a 40 per cent error in their couplings.

We have always stressed that SU_6 invariance is a non-relativistic concept, because very substantial difficulties are found in extending it into the relativistic domain. We cannot adequately describe here the generalizations and additional assumptions, the successes and failures of this research; instead we refer to recent reviews.[S 17, S 18]

CHAPTER 2

The Basic Ideas of Current Algebras

2.1. Current Commutators

In Section 1.2 we have summarized our knowledge about the isospin group by two statements:

(a) The isospin group is a symmetry group of strong interactions.

(b) The charges of the isovector part of the electromagnetic current and the strangeness conserving part of the weak current satisfy an SU_2 algebra (CVC).

In Sections 1.3 and 1.4 we gave a rough outline of theories which generalize statement (a) by introducing approximate higher symmetry groups. The generalization of statement (b) will lead to current algebras as proposed by Gell-Mann.[164]

We accept that strong interactions do not have a larger symmetry than the isospin group, but this does not exclude the possibility of constructing an algebra of observable operators larger than the SU_2 algebra (b). In the isospin algebra only certain parts of the weak currents and charges are used; we might ask whether there exists a larger algebra containing all the electromagnetic and weak charges. Of course we are looking for a finite algebra, containing not too many operators, and we are not interested in the infinite algebra generated by all observables, if one naturally accepts that the commutator of two observables is again an observable.

For the moment we confine ourselves to the vector currents observed in electromagnetic and weak interactions and leave the axial vector currents for later. We form their charges $Q^i = \int j_0^i(x)(dx)^3$, for instance Q^{II} with the strangeness changing weak vector current $j_\mu^{\mathrm{II}}(x)$, and find that some of them are time-dependent, for instance $dQ^{\mathrm{II}}/dt \neq 0$, since strong interactions are not invariant under a change of strangeness: $m_K \neq m_\pi$. We want to make assumptions about the commutators of the vector charges; if these are time-dependent we will confine ourselves to commutators at equal times.

The commutators of the strangeness conserving charges I^i and Y with each other are known through the isospin algebra and the isoscalar nature of the hypercharge Y. The commutators of the strangeness changing charges Q^{II} and $(Q^{II})^+$ with I^i and Y are known from the quantum numbers of j_μ^{II} (see equation (1.13)). Only the commutators of two strangeness changing charges are unknown. We will assume that they close to an SU_3 algebra, as described in Section 1.3, and we will also introduce Cabibbo's normalization.[S 12] We represent the physical weak currents from (1.13) like the SU_3 model currents in (1.19), (1.20), (1.21):

Vector part   Axial vector part

$$J_\mu^I = \cos \vartheta\,(j_\mu^1 + ij_\mu^2) + \cos \vartheta\,(\bar{j}_\mu^1 + i\bar{j}_\mu^2) \tag{2.1}$$

$$J_\mu^{II} = \sin \vartheta\,(j_\mu^4 + ij_\mu^5) + \sin \vartheta\,(\bar{j}_\mu^4 + i\bar{j}_\mu^5) \tag{2.2}$$

$$j_\mu^{el} = j_\mu^3 + \frac{1}{\sqrt{3}}j_\mu^8. \tag{2.3}$$

We form their charges $Q^i(t) = \int j_0^i(\bar{x},t)(d\bar{x})^3$ and now we postulate

$$[Q^i(t),Q^j(t)] = if^{ijk}Q^k(t). \tag{2.4}$$

It is important to stress the difference between the model relation (1.22) and the physical assumption (2.4). In the model of SU_3 symmetry the electromagnetic and weak currents were assigned to an octet by Cabibbo's theory; their charges, time-independent in the symmetry model, also form an octet. This octet, containing the isospin generators, is the octet of Lie generators in SU_3, so (1.22) follows. Here we deal with the full physical strong interactions, which have only isospin symmetry. We define the charges in (2.4) from physical currents. Large classes of their matrix elements are, in principle, directly observable in electromagnetic and weak interactions (to lowest order in e or G) using the effective Hamiltonians (1.1) and (1.2). In extending the isospin algebra we postulate (2.4) as a set of exact basic relations, applying not to a model, but to physical reactions.

As an approximate symmetry, SU_3 has only a vague and unclear interpretation; the algebra (2.4) is defined precisely and unambiguously in terms of physical currents. Its introduction does not depend on the extent to which SU_3 is an invariance group or not; invariance is a property of the Hamiltonian in relation to the algebra (see Section 3.1).

The study of algebras will be useful even in cases where no acceptable symmetry limit exists, for instance with the $SU_3 \times SU_3$ algebra of Section 2.2, where a symmetry limit would require massless hadrons,[166] or with the SU_6 algebra of Section 7.1, where a satisfactory symmetry limit exists only in nonrelativistic models.

Having defined in (2.4) the algebra of the SU_3 Lie-generators we can introduce finite SU_3 transformations in the space of physical hadron states by a generalized integration. Like the generators $Q^i(t)$, these transformations will be time-dependent and they will be frame-dependent, because a charge is a frame-dependent concept if a current is not conserved. To see this we write the space-integration as an integration over a three-dimensional surface S in four-dimensional space:

$$Q^i(S) = \int (d\bar{x})^3 j_0^i(\bar{x},t) = \int d\sigma^\mu j_\mu^i(x) \text{ with } d\sigma^\mu = ((d\bar{x})^3, \bar{O}). \quad (2.5)$$

The difference of two charges in Lorentz-transformed frames can be written as an integral over the current-divergence in the enclosed space-time continuum (Fig. 3), if contributions from

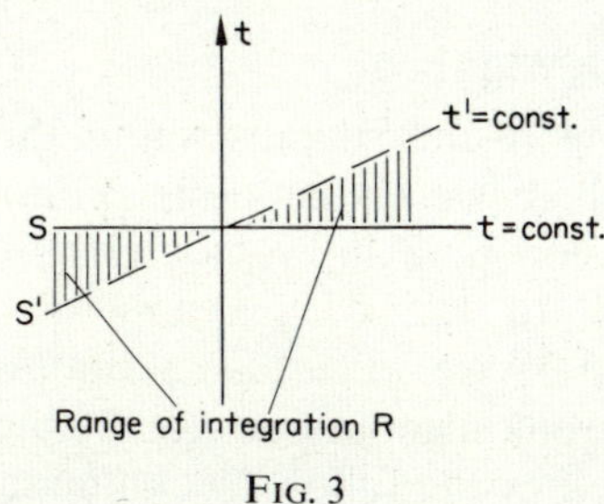

Fig. 3

infinite surfaces are neglected; this is always possible for matrix elements of wave-packet states.

$$Q^i(S) - Q^i(S') = \int_S d\sigma^\mu j^i_\mu(x) - \int_{S'} d\sigma'^\mu j^i_\mu(x) = \int_R (dx)^4 \partial^\mu j^i_\mu(x).$$

$$(2.6)$$

The charge is a scalar only if the current is conserved.

Three questions are now to be answered:

(a) Is the proposal of an exact, but time- and frame-dependent algebra at all consistent with relativistic invariance in the absence of a corresponding symmetry group? In Section 2.3 we will answer this question positively by showing that a simple relativistic quark model contains the SU_3 algebra and its extensions (Section 2.2) but without having SU_3 symmetry.

(b) Accepting that a current algebra is self-consistent, do the physical currents really satisfy it? Several sections will be devoted to the experimental confirmations of the SU_3 algebra and its extensions. In every practicable test so far, the algebra had to be combined with dynamical assumptions and approximations which have already been more or less successful in other applications. We find significant agreement in all cases where drastic approximations and disputable additional hypotheses can be avoided. There are not many cases of this kind (for instance the Adler–Weisberger relation[9,10,385–6]); sometimes we will find only rough agreement, but very often the predictions will be better than the approximations lead one to expect.

(c) Accepting that current algebras are correct, how are they connected to other dynamical theories? We will try to comment on this point in Chapter 10; until then, we shall take the commutators as working hypotheses to be tested.

2.2. The $SU_3 \times SU_3$ Algebra

Before discussing further details about the SU_3 algebra we wish to complete the basic assumptions of current algebras. This involves an extension to axial vector currents and charges of weak interactions. In the SU_3 symmetric model these are assigned to an octet; we take this assumption over into the framework of current algebras. Octet behaviour of the axial charges is expressed by the following commutation relations with the SU_3 generators:

$$[Q^i(t), \bar{Q}^j(t)] = if^{ijk}\bar{Q}^k(t)$$

$$\text{where} \quad \bar{Q}^j(t) = \int \bar{J}^j_0(\bar{x},t)\,(dx)^3. \qquad (2.7)$$

We take these relations as our second basic postulate to be tested. Next we consider the commutator of two axial charges. Gell-Mann and Ne'eman[168] proposed that the total hadronic weak charge $Q^{W,h}(t) = \int J_0^{W,h}(\bar{x},t)\,(dx)^3$ should generate an SU_2 algebra together with its hermitean adjoint, as does the total leptonic weak charge $Q^{W,l}(t) = \int J_0^{W,l}(\bar{x},t)\,(dx)^3$. By applying the usual equal-time anticommutation relations for spinor fields this can easily be checked from (1.6). We can also find the required normalization; if we want to have

$$[W_s^i, W_s^j] = ie_{ijk}W_s^k \tag{2.8}$$

we have to introduce:

$$W_s^1 = \frac{1}{4}(Q^{W,s}(t) + [Q^{W,s}(t)]^+),$$

$$W_s^2 = \frac{1}{4i}(Q^{W,s}(t) - [Q^{W,s}(t)]^+),$$

$$W_s^3 = \frac{1}{8}[Q^{W,s}(t), [Q^{W,s}(t)]^+], \tag{2.9}$$

where s stands for h or l: hadronic or leptonic charges.

The postulate (2.8) remains unchanged under unitary transformations

$$V_h^i = UW_h^iU^{-1}: \quad [V_h^i, V_h^j] = ie_{ijk}V_h^k. \tag{2.10}$$

For U we use an SU_3 transformation such that

$$UQ^{W,h}(t)U^{-1} = [Q^1(t) + iQ^2(t)] + [\bar{Q}^1(t) + i\bar{Q}^2(t)]; \tag{2.11}$$

it is simply a rotation in the Cartan diagram of SU_3 (Fig. 2, p. 13). To construct this transformation and to apply the following arguments it is essential that we have the same angle ϑ for vector and axial vector charges in $Q^{W,h}(t)$:

$$Q^{W,h}(t) = \cos\vartheta[(Q^1(t) + iQ^2(t)) + (\bar{Q}^1(t) + i\bar{Q}^2(t))]$$

$$+ \sin\vartheta[(Q^4(t) + iQ^5(t)) + (\bar{Q}^4(t) + i\bar{Q}^5(t))]. \tag{2.12}$$

With a suitably chosen transformation U in (2.11) we have

$$V_h^1 = \tfrac{1}{2}(Q^1(t) + \bar{Q}^1(t)),$$

$$V_h^2 = \tfrac{1}{2}(Q^2(t) + \bar{Q}^2(t)),$$

$$V_h^3 = \frac{1}{4i}([Q^1(t), Q^2(t)] + [Q^1(t), \bar{Q}^2(t)] + [\bar{Q}^1(t), Q^2(t)] + [\bar{Q}^1(t), \bar{Q}^2(t)])$$

$$= \tfrac{1}{2}(\tfrac{1}{2}Q^3(t) + \frac{1}{2i}[\bar{Q}^1(t), \bar{Q}^2(t)] + \bar{Q}^3(t)), \tag{2.13}$$

using (2.4) and (2.7). The simplest consistent solution would be to have $V_h^3 = \tfrac{1}{2}(Q^3(t) + \bar{Q}^3(t))$, this would imply the postulate

$$[\bar{Q}^i(t), \bar{Q}^j(t)] = ie_{ijk}Q^k(t) \quad (i, j, k = 1,2,3). \tag{2.14}$$

The precise reason why hadronic and leptonic weak charges should both generate an SU_2 algebra is not known at present, but by the non-linear relations (2.8) and (2.14) this principle offers us a model-independent definition of universality for weak couplings. If we want to speak about universal coupling constants e and G we must have a prescription for fixing the scale of the currents in the Hamiltonians (1.1) and (1.2). This scale can be fixed either by the existence of conserved charges or by non-linear conditions on the currents such as commutators. Before the theory of current commutators was proposed, the concept of universality was based on Lagrangian models with bare fields. A certain consistency could be achieved in the case of conserved currents (CVC), but the concept was unclear with the non-conserved axial vector currents.

If we now consider a commutator of two general octet axial charges $[\bar{Q}^i(t), \bar{Q}^j(t)]$, we know from the Clebsch–Gordan series of SU_3 that as an antisymmetric product of two operators from the same octet, the commutator is a linear combination of octet, decuplet and antidecuplet operators. In the special cases of (2.14) we see that only octet operators occur; by suitable SU_3 transformations on (2.14) we can convince ourselves that there are no decuplet and antidecuplet components in the other commutators either.[168] Therefore our third basic postulate for tests will be:

$$[\bar{Q}^i(t), \bar{Q}^j(t)] = if^{ijk}Q^k(t). \tag{2.15}$$

The mathematical structure of (2.4), (2.7) and (2.15) is the algebra of $SU_3 \times SU_3$. The two commuting SU_3 algebras are given by $(Q^i(t) \pm \bar{Q}^i(t))$. The symmetry limit of the algebra would require massless hadrons;[166] this does not seem to be a good approximation to strong interaction physics.

Of the generators Q^1, Q^2, Q^4, Q^5, $Q^3 + (1/\sqrt{3})\,Q^8$; $\bar{Q}^1$, $\bar{Q}^2$, $\bar{Q}^4$, $\bar{Q}^5$ are directly observable in electromagnetic and weak reactions; in the limit of isospin symmetry we can construct Q^6, Q^7, $\bar{Q}^3$, $\bar{Q}^6$, $\bar{Q}^7$ and also separate Q^3 from Q^8; $\bar{Q}^8$ is directly accessible only in the SU_3 symmetry limit. It is not known whether there is any deeper significance behind this distinction.

In the SU_3 symmetry model not only the charges but also the currents are considered in an operator octet. In current algebras this would amount to postulating:

$$[Q^i(t), j^j_\mu(\bar{x},t)] = if^{ijk} j^k_\mu(\bar{x},t), \qquad (2.16)$$

$$[Q^i(t), \bar{j}^j_\mu(\bar{x},t)] = if^{ijk} \bar{j}^k_\mu(\bar{x},t). \qquad (2.17)$$

Relations (2.16) and (2.17) are more general than (2.4) and (2.7). The latter can be obtained from (2.16) and (2.17) by taking current time-components and integrating over the space. In this integration magnetic currents are lost $[M_\mu(x) = \partial_\nu T_{\mu\nu}(x)$ with $T_{\mu\nu}(x) = -T_{\nu\mu}(x)]$, so (2.16) and (2.17) are indeed stronger assumptions. Two further non-trivial extensions are:

$$[\bar{Q}^i(t), j^j_\mu(\bar{x},t)] = if^{ijk} \bar{j}^k_\mu(\bar{x},t), \qquad (2.18)$$

$$[\bar{Q}^i(t), \bar{j}^j_\mu(\bar{x},t)] = if^{ijk} j^k_\mu(\bar{x},t). \qquad (2.19)$$

In most of the tests we will require the stronger postulates (2.16)–(2.19).

2.3. A Quark Model for the $SU_3 \times SU_3$ Algebra

First of all we have to show that current algebras, proposed in a non-covariant form in Sections 2.1 and 2.2, are compatible with relativistic invariance of the theory. To show that there is no contradiction, it is sufficient to give a model which is relativistically invariant and contains current algebras. We choose a simple quark model,[164–65] described by a Lagrangian density $\mathscr{L}$

$$\mathscr{L} = i\bar{q}(x)\gamma_\mu \frac{\overleftrightarrow{\partial}}{\partial x_\mu} q(x) - m(\bar{p}(x)p(x)$$

$$+ \bar{n}(x)n(x)) - m'(\overline{\lambda(x)\lambda(x)}). \quad (2.20)$$

Here $q(x)$ is a triplet of free fermion fields, described by Dirac spinors.

As vector and axial vector currents we take

$$j_\mu^i(x) = \bar{q}(x)\gamma_\mu \tfrac{1}{2}\lambda_i q(x). \quad (2.21)$$

$$\bar{j}_\mu^i(x) = \bar{q}(x)\gamma^5\gamma_\mu \tfrac{1}{2}\lambda_i q(x). \quad (2.22)$$

By simple commutation† of quark field operators according to the usual anticommutators

$$\{\Psi_\alpha^+(x),\Psi_\beta(y)\}_+ = \delta_{\alpha\beta}\delta^3(x-y) \quad \text{for} \quad x_0 = y_0 \quad (2.23)$$

(α,β stand for spin and SU_3 indices) we deduce for $x_0 = y_0$

$$[j_0^i(x),j_\mu^j(y)] = if^{ijk}j_\mu^k(x)\delta^3(x-y), \quad (2.24)$$

$$[j_0^i(x),\bar{j}_\mu^j(y)] = if^{ijk}\bar{j}_\mu^k(x)\delta^3(x-y), \quad (2.25)$$

$$[\bar{j}_0^i(x),j_\mu^j(y)] = if^{ijk}\bar{j}_\mu^k(x)\delta^3(x-y), \quad (2.26)$$

$$[\bar{j}_0^i(x),\bar{j}_\mu^j(y)] = if^{ijk}j_\mu^k(x)\delta^3(x-y). \quad (2.27)$$

From (2.24)–(2.27) we can deduce the charge-current commutation relations (2.16)–(2.19) by space-integration over x and, with current time-components ($\mu = 0$), perform another space integration to derive the charge-commutators (2.4), (2.7) and (2.15). The fact that we have taken different quark masses ($m' \neq m$) and so destroyed SU_3 symmetry, has not at all influenced the derivation. One may also add interactions of various kinds, but as long as the equal time commutators (2.23) and the structure of the currents are maintained [(2.21) and (2.22)], the current algebra remains valid for the model.

Different views are held about the importance of this quark model for the physical current algebra. Its success would not imply that quarks exist like particles; it has often been possible to abstract from special models statements of general validity.

†For Schwinger terms see Chapter 8.

On the other hand not all kinds of field theoretical models give rise to the same current algebra. If we enlarge the quark model by the addition of a fundamental triplet $B(x)$ of free spinless bosons, the vector current would be changed:

$$j_\mu^i(x) = \bar{q}(x)\gamma_\mu \tfrac{1}{2}\lambda_i q(x) + iB^+(x)\tfrac{1}{2}\lambda_i \overset{\leftrightarrow}{\partial}_\mu B(x) \qquad (2.28)$$

but the axial current would keep its form (2.22) because it is not possible to construct an axial vector current as a bilinear form in $B(x)$. Consequently the commutator of two axial vector charges in (2.15) would only give the quark part of the vector charge and violate the $SU_3 \times SU_3$ algebra. A singlet boson, however, would not violate the algebra because it does not contribute to either the vector current or the axial current.

For the details of the proposed models we refer to the literature; the SU_3 algebra by itself does not seem to have much power to discriminate between different models, and real distinctions require the use of axial charge commutators or of higher algebras, as discussed in Chapter 7.

Models of the SU_3 algebra, including the construction of D-currents: (265), (268), (276).†

Construction of an axial vector current and its algebra by trilinear expressions in pseudoscalar fields: (237).

Tests of quark models (charge assignment) with extended algebras: (226), (292,–94).

Maximal chiral group of the quark model: (299).

Review of model tests by current algebras: (294).

Further work on models for current algebras‡ is listed in Chapter 8, where we will discuss the studies of Schwinger terms.

2.4. Lepton Scattering Sum Rules

The most direct way of testing a commutator is the insertion of a complete set of intermediate states into its matrix elements. Using invariant normalization

$$\langle a|b\rangle = (2\pi)^3\delta^3(p^a - p^b)2E_p \qquad (2.29)$$

we obtain the sum rule

---

†Numbered references under Bibliography on Current Algebra at end of book.

‡Recent contributions: (422), (456), (463).

$$\langle a|[Q^i, Q^j]|b\rangle = \sum_n \langle a|Q^i|n\rangle\langle n|Q^j|b\rangle \over (2\pi)^3 2E_n} - {\langle a|Q^j|n\rangle\langle n|Q^i|b\rangle \over (2\pi)^3 2E_n}$$

$$= if^{ijk}\langle a|Q^k|b\rangle \quad (2.30)$$

$$\langle a|Q^k|b\rangle = \int (dx)^3\langle a|j_0^k(x)|b\rangle$$

$$= \int (dx)^3 \exp i(p^a - p^b)x . \langle a|j_0^k(0)|b\rangle$$

$$= (2\pi)^3 . \delta^3(p^a - p^b)\,\langle a|j_0^k(0)|b\rangle. \quad (2.31)$$

Inserting (2.31) into (2.30) gives finally

$$\sum_n{}' {\langle a|j_0^i(0)|n\rangle\langle n|j_0^j(0)|b\rangle \over 2E_n} - {\langle a|j_0^j(0)|n\rangle\langle n|j_0^i(0)|b\rangle \over 2E_n}$$

$$= if^{ijk}\langle a|j_0^k(0)|b\rangle, \quad (2.32)$$

where $\sum_n{}'$ denotes a restricted summation: only over states with equal spatial momentum. Sum rules like (2.32) will be called direct sum rules. If weak and electromagnetic reactions of many-particle states (contained in $\sum_n |n\rangle\langle n|$) were known experimentally, the sum rules could be directly compared with experiment. With some skill it is possible to rewrite part of them into a form containing not particular matrix elements, but differential cross sections for lepton reactions. As an example we present one of Adler's sum rules for neutrino reactions[10] as an exact but as yet untestable consequence of current algebras.

We take the commutator of two axial charges between protons of momentum $\bar{p}$

$$\langle P(\bar{p})|[\bar{Q}^1 + i\bar{Q}^2, \bar{Q}^1 - i\bar{Q}^2]|P(\bar{p})\rangle = 2\langle P(\bar{p})|Q^3|P(\bar{p})\rangle$$

$$= \langle P(\bar{p})|P(\bar{p})\rangle. \quad (2.33)$$

Explicitly we separate the contribution of the neutron intermediate state from the continuum, using the notation (1.7) and neglecting the form factors $G_{2,PN}^w$ and $G_{3,PN}^w$ because the contribution of their kinematic forms is of the order $(m_P - m_N)$, and for the same reason neglecting the dependence of $G_{1,PN}^w$ on the momentum transfer. We introduce the axial vector coupling constant g_A by $G_{1,PN}^w(0)/\cos\theta = g_A \approx 1.18$. g_A is the quotient of the measured axial and vector coupling constants in β decay.

28

$$\langle P(\bar{p}) \mid (\bar{j}^1_\mu(0) + i\bar{j}^2_\mu(0) \mid N(\bar{p}) \rangle \approx \bar{u}_P(\bar{p})\gamma^5\gamma_\mu g_A u_N(\bar{p}) \quad (2.34)$$

consequently

$$2E_P = \sum_{\text{pol}} \bar{u}_P(\bar{p})\gamma^5\gamma_0 \frac{u_N(\bar{p})\bar{u}_N(\bar{p})\gamma^5\gamma_0 u_P(\bar{p})}{2E_N}|g_A|^2$$

$$+ \sum_{n \neq N}' \left\{ |\langle P(\bar{p})|(\bar{j}^1_0 + i\bar{j}^2_0)|n(\bar{p})\rangle|^2 \frac{1}{2E_n} \right.$$

$$\left. - |\langle P(\bar{p})|(\bar{j}^1_0 - i\bar{j}^2_0)|n(\bar{p})\rangle|^2 \frac{1}{2E_n} \right\} \quad (2.35)$$

In the neutron term we sum over the spin and obtain

$$\sum_{\text{pol}} \bar{u}_P(\bar{p})\gamma^5\gamma_0 u_N(\bar{p})\bar{u}_N(\bar{p})\gamma^5\gamma_0 u_P(\bar{p})$$

$$= \bar{u}_P(\bar{p})\gamma^5\gamma_0(\gamma p^N + m_N)\gamma^5\gamma_0 u_P(\bar{p})$$

$$= \bar{u}_P(\bar{p})\gamma_0(\gamma p^N - m_N)\gamma_0 u_P(\bar{p}) = \bar{u}_P(\bar{p})(\gamma_0 E^N + \bar{\gamma}\bar{p} - m_N)u_P(\bar{p})$$

$$= 2\bar{u}_P(\bar{p})(\bar{\gamma}\bar{p})u_P(\bar{p}) \quad \text{(in the approximation } m_N = m_P \Rightarrow E^N = E^P)$$

$$= 4E_P^2\left(1 - \left(\frac{m_P}{E_P}\right)^2\right). \quad (2.36)$$

In the continuum we introduce the divergences $\bar{D}^{(\pm)} = \partial^\mu(\bar{j}^1_\mu \pm i\bar{j}^2_\mu)$ and integrate over all internal degrees of freedom of the intermediate states, leaving only the integration over the total mass to be done. We define the relativistic scalars

$$H^{(\pm)}(W,\bar{p}) = \sum_n \delta(m_n - W)|\langle P(\bar{p})|\bar{D}^{(\pm)}(0)|n(\bar{p})\rangle|^2 \quad (2.37)$$

and obtain from (2.35)†

$$1 = \left(1 - \left(\frac{m_P}{E_P}\right)^2\right)|g_A|^2 + \int_{(m_P+m_\pi)}^{\infty} \left(\frac{dW}{2E_W \cdot 2E_P \cdot (E_W - E_P)^2}\right.$$

$$\left. [H^{(+)}(W,\bar{p}) - H^{(-)}(W,\bar{p})]\right). \quad (2.38)$$

---

†We rewrite $\langle P|j_0|n\rangle = \dfrac{\langle P|\partial^\mu j_\mu|n\rangle}{i(E_P - E_n)}$, because the spatial divergence does not contribute between states of equal momentum.

The functions $H(W,\bar{p})$ are relativistic invariants because they contain squares of pseudoscalar matrix elements between proton states and states of definite mass W, all invariantly normalized. The proton momentum $\bar{p}$ specifies only the momentum transfer $q^2 = (E_W - E_P)^2 = (\sqrt{(W^2 + \bar{p}^2)} - \sqrt{(m_P^2 + \bar{p}^2)})^2$ in $H(W,\bar{p})$, which is dependent on the mass of the intermediate state; only in the limit $\bar{p} \to \infty$ is it asymptotically independent: $q^2 \approx (1/2\bar{p}\,(m_P^2 - W^2))^2 \to 0$.

We take the limit $\bar{p} \to \infty$ and make the assumption that we can take it at individual intermediate-state contributions and integrate their limits; this implies an interchange of the limit $\bar{p} \to \infty$ with the limiting process of summation over an infinite intermediate state spectrum. The significance of this assumption will be discussed in Chapter 4. Denoting $\lim H(W,\bar{p})$ simply by $H(W)$ we end up with a covariant equation:

$$1 = |g_A|^2 + \int_{(m_P + m_\pi)}^{\infty} \frac{dW}{(W^2 - m_P^2)^2} [H^+(W) - H^-(W)]. \quad (2.39)$$

$H(W)$ has to be taken at zero momentum transfer between the proton and the continuum states; because it is relativistically invariant, the state of motion of the proton has no influence on its value and we may take the proton at rest provided we keep zero momentum transfer.

To connect $H(W)$ with observable cross sections we first derive a theorem of Adler.[6] The differential cross section for lepton forward scattering in an inelastic reaction $\nu + P \to l + (n)$ with $m_n \neq m_P$ depends only on the current divergence, if the lepton mass is neglected. The matrix element is proportional to

$$\frac{1}{\sqrt{2}} \bar{u}_l \gamma_\mu (1 + \gamma^5) u_\nu \langle n | J_\mu^{W,h} | P \rangle. \quad (2.40)$$

Squaring and summing over the lepton spin gives

$$4 \langle P | (J_\lambda^{W,h})^+ | n \rangle \langle n | J_\mu^{W,h} | P \rangle \{ p_\mu^l p_\lambda^\nu + p_\mu^\nu p_\lambda^l - g_{\mu\lambda}(p^\nu p^l) \} \quad (2.41)$$

neglecting the lepton mass. For forward scattering p^l, p^ν and $(p^p - p^n) = (p^l - p^\nu)$ are parallel vectors of modulus zero.

30

The product (2.41) can then be expressed as

$$8|\langle n|(\partial^\mu J_\mu^{W,h})|P\rangle|^2 \frac{E^l E^\nu}{(E^\nu - E^l)^2} \stackrel{\text{def}}{=} 8|M_\nu^n|^2 \frac{E^l E^\nu}{(E^\nu - E^l)^2}. \quad (2.42)$$

With a strangeness conserving reaction the divergence of the vector current is zero in the limit of isopin symmetry (see Section 1.2), so only the axial vector current contributes. From the quantum numbers of the weak current we see that neutrino absorption is always coupled to the $\Delta Q = +1$ part of $J_\mu^{W,h}$. Therefore $H^{(-)} = \sum_n \delta(m_n - W)|M_\nu^n|^2$ and $H^{(+)}$ is the corresponding quantity for antineutrino reactions.

For $E^\nu > (m_n^2 - m_P^2)/2m_P$ the quantity $|M_\nu^n|^2$ is expressible as a differential cross-section.

$$\frac{d\sigma_n^\nu}{d\Omega_l} = \int \frac{(G\cos\vartheta)^2 8|M_\nu^n|^2 E^l \cdot E^\nu}{(2\pi)^2 16 E^n \cdot m_P \cdot E^l \cdot E^\nu \cdot (E^\nu - E^l)^2} (E^l)^2 dE^l \cdot d^3\bar{p}^n$$
$$\delta^3(\bar{p}^n + \bar{p}^l - \bar{p}^\nu) \cdot \delta(E^n + E^l - E^\nu - m_P)$$

$$= \int \frac{(G\cos\vartheta)^2 |M_\nu^n|^2 (E^l)^2 dE^l}{(2\pi)^2 \cdot 2E^n \cdot m_P (E^\nu - E^l)^2} \delta(\sqrt{(m_n^2 + (E^l - E^\nu)^2)}$$
$$+ E^l - E^\nu - m_P)$$

$$= \frac{(G\cos\vartheta)^2 |M_\nu^n|^2 (E^l)^2}{(2\pi)^2 2m_P^2 (E^\nu - E^l)^2}$$

$$= \frac{(G\cos\vartheta)^2 |M_\nu^n|^2 (m_P^2 + 2m_P E^\nu - m_n^2)^2}{(2\pi)^2 2m_P^2 (m_n^2 - m_P^2)^2}. \quad (2.43)$$

We substitute (2.43) into (2.39) and obtain: $1 = |g_A|^2 +$

$$+ \int_{(m_P + m_\pi)}^{\infty} \frac{(2\pi)^2 2m_P^2 [(d\sigma^{\bar{\nu}}/d\Omega_l)(E^{\bar{\nu}}, W) - (d\sigma^\nu/d\Omega_l)(E^\nu, W)]}{(G\cos\vartheta)^2 (m_P^2 + 2m_P E^\nu - W^2)^2} dW \quad (2.44)$$

where the neutrino or antineutrino energy $E^\nu(E^{\bar{\nu}})$ is a free parameter. For $W^2 > 2m_P E^\nu + m_P^2$ the integrand has to be continued analytically below threshold. $d\sigma^\nu/d\Omega_l$ is the differential forward cross-section for all kinds of inelastic reactions $\nu + P \rightarrow (n) + $ lepton, where (n) is any hadron system with strangeness zero excluding the nucleons. This quantity is not yet sufficiently well known experimentally to test the sum rule (2.44).

A similar sum rule can be constructed for strangeness changing reactions [10]. With our present knowledge on inelastic reactions we cannot test the direct sum rules. But there exist methods to approximate them by known quantities with the help of dispersion relations; these will be discussed in Chapters 4, 5, 6 and 9.

2.5. Some Remarks on the Mathematical Description of Currents and Charges

We have introduced charges and currents physically; to determine the matrix element of some current between asymptotically free states of strong interactions we have to think of a possible first-order weak or electromagnetic reaction producing a transition between the states of interest and have to use the Hamiltonians (1.1) and (1.2) to determine the current matrix element in lowest order of e or G. (The extension to higher orders is certainly not trivial.) We assume that the asymptotically free states of strong interactions form a complete basis in the space of hadrons; thus all matrix elements of the currents are defined.

This physical approach to currents and charges does not solve the problem of their exact mathematical description. In particular the non-conserved charges $Q(t)$ are not easily accommodated among the conventional classes of operators. Fabri and Picasso[122] have shown: If $Q(t) = \int (dx)^3 j_0(\bar{x},t)$ exists (as a weak limit) and is defined on the vacuum state, it must annihilate it.

Proof. $Q(t)$ is translationally invariant; so the same is true for $|Q\rangle = Q(t)|0\rangle$. Then

$$\langle Q|Q\rangle = \langle Q|Q(t)|0\rangle = \int (dx)^3 \langle Q|j_0(\bar{x},t)|0\rangle$$

$$= \int (dx)^3 \langle Q|j_0(0,t)|0\rangle \quad (2.45)$$

is finite only if it is zero.

Coleman[105] has shown that $Q|0\rangle = 0$ implies that Q is a constant of the motion.

An even stronger result holds:[123] Q cannot be defined as a self-adjoint operator in the Hilbert space, unless there is an exact symmetry: $\partial^\mu j_\mu(x) = 0$. *Proof.* If Q is self-adjoint, $e^{i\lambda Q}$ exists for all states and because of translational invariance $e^{i\lambda Q}|0\rangle = e^{i\lambda q}|0\rangle$ with a real number q. Then by Stone's theorem $Q|0\rangle = q|0\rangle$. Since $\langle 0|j_\mu(x)|0\rangle = 0$ by Lorentz invariance, it follows that $q = 0$ and we can again use Coleman's theorem[105] to infer that Q has to be a constant of the motion.

There are two ways to introduce non-conserved charges:

(a) an extension of the Hilbert space into a larger topological vector space (213A);

(b) a restriction to matrix elements between suitably localized states,[339,350] which is certainly acceptable since all experiments are performed in macroscropically finite regions.

There are further problems if we start to specify equal-time commutators rigorously; the formulation

$$\sum_n \left(\frac{\langle a|Q^i|n\rangle\langle n|Q^j|b\rangle}{(2\pi)^3 2E_n} - \frac{\langle a|Q^j|n\rangle\langle n|Q^i|b\rangle}{(2\pi)^3 2E_n} \right) = if^{ijk}\langle a|Q^k|b\rangle \quad (2.46)$$

would be free from difficulties only if $\langle a|Q^iQ^j|b\rangle$ and $\langle a|Q^jQ^i|b\rangle$ did separately converge with a complete set of intermediate states. This is not the case, as we will see in the Adler–Weisberger relation (Section 4.2). So we need additional prescriptions for introducing intermediate states; practical success is achieved with dispersion theoretic concepts.

It is beyond the level of this book to provide a rigorous solution of these problems. We will remain on the somewhat intuitive and experimental level of the original papers and refer the reader to some selected literature for a mathematically more elaborate treatment.

Difficulties in the definition of charges, in particular with non-conserved ones: references (122–3), (213A), (339), (350), (369).

Treatment of current algebras on the level of the LSZ formalism, including the dispersion theory of current algebras: reference (350).

Treatment of current algebras in the framework of the Haag–Kastler algebraic theory of local observables: reference (339).

CHAPTER 3

Approximate SU₃ Symmetry in Current Algebra Theory

3.1. The Description of an Approximate Symmetry

In Section 2.1 we constructed the SU_3 algebra from physical currents without reference to the extent that SU_3 symmetry is broken; we went further by developing the $SU_3 \times SU_3$ algebra which does not correspond to any familiar symmetry group. It is nevertheless useful to develop the concept of an approximate symmetry and its consequences, but this time from the current algebra point of view; we will see that this specification of the approximate-symmetry concept is supported by experimental results.

In current algebra theory, results of approximate-symmetry models can be reproduced in two situations. The first situation arises if the Hamiltonian $\mathcal{H}$ can be split into a part $\mathcal{H}_0$ commuting with the generators and another part $\mathcal{H}'$ which may in some sense be regarded as small. Such a situation may be true with respect to the SU_3 algebra (2.4)

$$H = H_0 + \lambda H' \quad \text{with} \quad [Q^i, H_0] = 0 \quad \text{and} \quad \lambda \ll 1. \tag{3.1}$$

The constant λ in (3.1) is not necessarily a fundamental coupling constant of SU_3 violations, we use it only as a scaling parameter to compare orders $0(\lambda^n)$.

Many deductions will start from the identity

$$-\left\langle a \middle| i\frac{dQ}{dt} \middle| b \right\rangle = (E_b - E_a)\langle a|Q|b\rangle = \langle a|[H, Q]|b\rangle$$
$$= \lambda\langle a|[H',Q]|b\rangle. \tag{3.2}$$

We see that $\langle a|Q|b\rangle$ can be of order 1 only if $(E_a - E_b)$ and consequently $(m_a - m_b)$ is of order $0(\lambda)$; (remember $\bar{p}_a = \bar{p}_b$). If

$(m_a - m_b)$ and consequently $(E_a - E_b)$ are of order 1, then $\langle a|Q|b \rangle$ can only be of order $0(\lambda)$.

Next we consider a commutator between states whose masses differ only in order $0(\lambda)$, substitute a complete set of intermediate states:

$$\sum_n = \sum_N {}^{\Delta m = 0(\lambda)} + \sum_{N'}$$

and separate the states with masses nearby

$$(\Delta m = 0(\lambda) \text{ in } \sum_N) \quad \text{from the rest} \quad (\sum_{N'})$$

which give only contributions in $0(\lambda^2)$. In this section we will take unit normalization for states, denoting them by $\langle\!\langle a| \ldots |b \rangle\!\rangle$:

$$if^{ijk}\langle\!\langle a|Q^k|b \rangle\!\rangle = \sum_N \{\langle\!\langle a|Q^i|N \rangle\!\rangle \langle\!\langle N|Q^j|b \rangle\!\rangle$$
$$- \langle\!\langle a|Q^j|N \rangle\!\rangle \langle\!\langle N|Q^i|b \rangle\!\rangle\} + 0(\lambda^2) \qquad (3.3)$$

For small λ the number of states nearby in mass is not arbitrary; up to small corrections they must provide a representation of the algebra, i.e. the subspace of states $|N \rangle\!\rangle$ must allow the construction of finite matrices $\langle\!\langle N_a|Q^i|N_b \rangle\!\rangle$ which up to small corrections satisfy the group algebra of SU$_3$. This is only possible if these sets of states form approximate SU$_3$ multiplets, since a representation of the algebra in a finite dimensional vector space always allows the construction of a group representation. Taking such a multiplet of (3.3) at rest, we see that its members must have the same spin and parity because the charges are unaffected by rotations and by space inversion. It is not necessary for the approximate SU$_3$ multiplets to be irreducible; if there are two multiplets of equal spin and parity and their average mass difference is of the order of the mass splittings within a multiplet, then mixing will occur. This is the case with the $\omega - \phi$ and to a smaller extent with the $\eta - X^0$ mesons. The physical particle-states do not approximately coincide with the octet or singlet basis vectors.

These arguments are strictly applicable only to discrete states because in the continuum energy differences cannot strictly be classified in orders of λ. This corresponds to the empirical result that not much of SU$_3$ symmetry is seen in scattering experiments except for resonances where single-particle approximations become meaningful again.

35

Apart from establishing multiplets and approximate selection rules the main use of symmetry groups has been in classifying observables as generalized tensors of the group (see Section 1.2). Every class of SU_3 tensors has characteristic commutation relations with the generators

$$[Q^i, T^j] = c^{ijk} T^k \quad \text{(for octets } c^{ijk} = if^{ijk}). \tag{3.4}$$

We take (3.4) between the states of some multiplet and separate the states of the same multiplet ($\underset{N}{\Sigma}$) from others ($\underset{N'}{\Sigma}$),

$$\underset{N}{\Sigma} \{\langle\!\langle a|Q^i|N\rangle\!\rangle \langle\!\langle N|T^i|b\rangle\!\rangle - \langle\!\langle a|T^j|N\rangle\!\rangle \langle\!\langle N|Q^i|b\rangle\!\rangle\}$$

$$+ \underset{N'}{\Sigma} \{\underset{0(\lambda)}{\langle\!\langle a|Q^i|N'\rangle\!\rangle} \quad \underset{0(\lambda) \text{ if } T^j = Q^j}{\langle\!\langle N'|T^j|b\rangle\!\rangle} - \underset{0(\lambda) \text{ if } T^j = Q^j}{\langle\!\langle a|T^j|N'\rangle\!\rangle} \quad \underset{0(\lambda)}{\langle\!\langle N'|Q^i|b\rangle\!\rangle}\}$$

$$= c^{ijk} \langle\!\langle a|T^k|b\rangle\!\rangle. \tag{3.5}$$

In general the matrix element $\langle\!\langle a|T^k|b\rangle\!\rangle$ will receive corrections of order λ from many-particle intermediate states, which will vanish only in the symmetry limit, where $\langle\!\langle a|T^k|b\rangle\!\rangle$ can be determined from (3.5) by the Wigner–Eckart theorem in terms of a few scaling constants. So we expect in general corrections of order $0(\lambda)$ to the symmetry values of $\langle\!\langle a|T^k|b\rangle\!\rangle$.

But if the operators T^k are themselves the generators: $T^k = Q^k$, then the many-particle contributions in $\underset{N'}{\Sigma}$ are of order $0(\lambda^2)$ and one would expect that the corrections to $\langle\!\langle a|Q^k|b\rangle\!\rangle$ are also of second order only. For a proof we choose commutators of opposite charges: $[Q^4 + iQ^5, Q^4 - iQ^5] = Q^3 + \sqrt{3}Q^8$ and $[Q^6 + iQ^7, Q^6 - iQ^7] = -Q^3 + \sqrt{3}Q^8$, where the right-hand side is a conserved charge whose matrix elements are unaffected by SU_3 breaking. We choose the states $\langle\!\langle a| \ldots |b\rangle\!\rangle$ such that there will only be one term in the multiplet contributions $\underset{N}{\Sigma}$, for instance with proton and Σ^+:

$$\langle\!\langle P|Q^6 + iQ^7|\Sigma^+\rangle\!\rangle \langle\!\langle \Sigma^+|Q^6 - iQ^7|P\rangle\!\rangle + 0(\lambda^2)$$

$$= \langle\!\langle P|-Q^3 + \sqrt{3}Q^8|P\rangle\!\rangle$$

$$|\langle\!\langle P|Q^6 + iQ^7|\Sigma^+\rangle\!\rangle|^2 + 0(\lambda^2) = 1. \tag{3.6}$$

So the symmetry-breaking correction to $\langle\!\langle P|Q^6 + iQ^7|\Sigma^+\rangle\!\rangle$ is of second order only; by isospin invariance this is true for all nucleon-sigma matrix elements of SU$_3$ generators, and now we can proceed to nucleon-lambda matrix elements etc. We can summarize the result by saying that the matrix elements of the generators Q^i within an irreducible (!) SU$_3$ multiplet are not affected by first-order SU$_3$ violations. This is the theorem of Ademollo and Gatto,[S 15] rederived from current algebras by Furlan, Lannoy, Rossetti and Segrè.[(149)] It explains why Cabibbo's theory[S 12] gives unexpectedly good predictions for weak vector currents and why Cabibbo's angle θ can be determined quite reliably[S 14] by comparing matrix elements of weak vector currents in the SU$_3$ symmetry model to real decay processes. No corresponding theorem holds for the axial vector currents because their charges are not generators of an approximate symmetry group. Their matrix elements are affected by first order SU$_3$ violations and are less suitable for determining θ.

Apart from the order estimates used to derive the Ademollo–Gatto theorem, the only important fact has been that in commutators like (3.5) only a finite set of intermediate states give appreciable contributions. Technically one often speaks of this set of states as saturating the commutators. For consistency this set must provide a representation of the algebra; along with the Wigner–Eckart theorem this is sufficient to give approximately the results of the corresponding symmetry model. Dominance of certain intermediate states can arise as a consequence of approximate symmetry (3.1), but it may also have quite different causes in special cases. Therefore with suitable particle multiplets this dominance may simulate an approximate symmetry group by reproducing its results. As will be discussed in Chapter 7, this may be the case with SU$_6$.

Symmetry models emerge as an approximation to the theory of current commutators, if only certain classes of intermediate states are kept; this has been gradually recognized in the literature: (27), (137), (282), (290), (358). As compared with symmetry models, current algebras contain a conceptual and a practical progress:

(a) Approximate saturation of commutators by finite sets of states may arise in special circumstances (for instance with certain single-particle states or in the rest system), but we are no longer burdened with the general implications of the corresponding symmetry model.

(b) Current algebras provide definite expressions for the corrections to symmetry models through sum rules; these allow us to understand general trends of the corrections and to make quantitative estimates, if the relevant data are known. Unfortunately this is not yet the case.

3.2. The Momentum Dependence of Direct Sum Rules

Among the significant successes of SU_3 are the Gell-Mann Okubo mass formulas, derived from the hypothesis that the symmetry breaking part $\mathscr{H}'$ of the Hamiltonian in (3.1) is a member of an octet. In current algebra language this would mean: $H' = N^8$ with

$$[Q^i, N^j] = if^{ijk}N^k. \tag{3.7}$$

The octet partners of H' include the time-derivatives of the strangeness changing charges, for example of $(Q^4 + iQ^5) \stackrel{\text{def}}{=} Q^{K+}$:

$$-i\frac{dQ^{K+}}{dt} = [H, Q^{K+}] = \lambda[N^8, Q^{K+}] = \frac{\sqrt{3}}{2}\lambda N^{K+}$$

$$= \frac{\sqrt{3}}{2}\lambda(N^4 + iN^5). \tag{3.8}$$

As a consequence of (3.7) we have:

$$[Q^{K+}, N^{K+}] = 0 = [Q^{K+}, [H, Q^{K+}]]. \tag{3.9}$$

We take (3.9) between $\langle\!\langle P|$ and $|\Xi^-\rangle\!\rangle$; as intermediate states we have Λ and Σ^0 and states outside the octet. According to the theorem of Ademollo and Gatto (Section 3.1) the matrix elements of Q^{K+} are

$$\langle\!\langle P|Q^{K+}|\Sigma^0\rangle\!\rangle = \frac{1}{\sqrt{2}} + 0(\lambda^2),$$

$$\langle\!\langle P|Q^{K+}|\Lambda\rangle\!\rangle = \sqrt{\frac{3}{2}} + 0(\lambda^2),$$

$$\langle\!\langle \Sigma^0|Q^{K+}|\Xi^-\rangle\!\rangle = \frac{1}{\sqrt{2}} + 0(\lambda^2),$$

$$\langle\!\langle \Lambda|Q^{K+}|\Xi^-\rangle\!\rangle = \sqrt{\frac{3}{2}} + 0(\lambda^2). \tag{3.10}$$

States outside the octet give contributions of order $0(\lambda)$ in every matrix element of Q^{K+}, so altogether their contributions are of order $0(\lambda^2)$. So we finally obtain from (3.9) and (3.10)

$$E_\Sigma + 3E_\Lambda - 2E_\Xi - 2E_P = 0(\lambda^2) \qquad (3.11)$$

and
$$E_\pi + 3E_\eta - 4E_K = 0(\lambda^2), \qquad (3.12)$$

if we take (3.9) between $\langle\!\langle K^+|$ and $|K^-\rangle\!\rangle$. In the rest frame we derive linear mass formulas, at infinite momentum quadratic ones: $E = \sqrt{(\bar{p}^2 + m^2)} \approx \bar{p} + m^2/2\bar{p}$. But linear and quadratic mass formulas differ only by second-order terms: $\Delta(m^2) = 2m\Delta m + (\Delta m)^2$; the corrections from non-octet states are of the same order and of course also dependent on the momentum of the states in the sum rule.

This momentum dependence is a consequence of the non-covariant formulation of the commutators in terms of space integrals. We will see in Chapter 4 that sum rules taken at different momenta are related to each other by shifting the surface of integration for a dispersion relation in a space of several complex variables. Given suitable analyticity properties, the sum rules are perfectly compatible with each other. A problem arises only in approximations: In which frames are the unknown many-particle contributions smallest so that they may be neglected?

Two special frames are distinguished from the rest for practical calculations: (a) the rest frame, by the fact that only states of equal spin contribute to the sum rules, and (b) the infinite momentum limit.

Charges connect states at equal spatial momentum, so the relativistic momentum transfer Δ^2 is the square of the energy difference. For any individual pair of states Δ^2 is positive and tends to zero at infinite momentum, independently of the masses. From the dispersion analysis of current form factors (Section 5.1), it is known that current matrix elements have singularities at positive Δ^2, so we cannot safely expect them to be small in their singularity region. Furthermore, at zero momentum transfer we can relate the matrix elements of the currents to observed quantities by using their nearest singularities (see Chapter 5), but not enough is known about multiparticle reactions, which determine the singularities at high momentum transfer.

The infinite momentum limit is only useful if we are allowed to sum the limits taken separately at individual intermediate states of definite masses. This means that we interchange the limiting process of summation over an infinite number of intermediate states with the limit $p \rightarrow \infty$:

$$\lim_{p \rightarrow \infty} \int dm f(m,p) = \int dm \lim_{p \rightarrow \infty} f(m,p). \qquad (3.13)$$

To understand the implications of this interchange, we consider disconnected intermediate states. Non-conserved charges applied to the vacuum will always connect it to other states:

$$\langle 0|Q|X \rangle \neq 0 \quad \text{if} \quad \frac{dQ}{dt} \neq 0. \qquad (3.14)$$

Consequently there will be disconnected contributions among the many-particle amplidutes of the form:

$$\langle p|Q|p; X \rangle = \langle p|p \rangle \langle 0|Q|X \rangle, \qquad (3.15)$$

graphically:

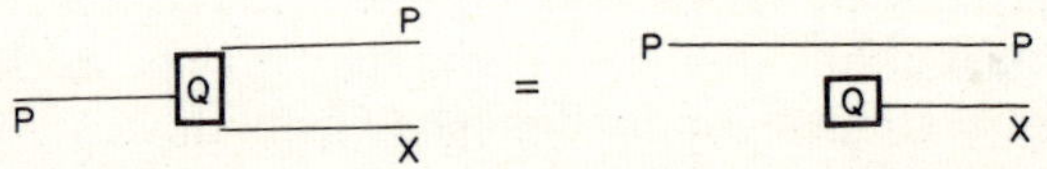

It is interesting to consider the invariant mass of the state

$$|p_i; X \rangle: M^2 = (E_p + E_X)^2 - \bar{p}_p^2 = M_p^2 + 2E_p E_X + E_X^2.$$

As $\bar{p}_p \rightarrow \infty$, M^2 will tend to infinity. At any given finite momentum disconnected configurations will occur among the intermediate states and may contribute to the commutator. If we take the infinite momentum limit at intermediate states of definite mass and sum over their contributions we will not include the disconnected states because they would correspond to an infinite mass. Evidently the interchange of limits in (3.13) is only allowed if the contribution of disconnected states vanishes asymptotically as $p_p \rightarrow \infty$.

40

It is not easy to see *a priori* when this will be the case. It will certainly depend on the operator Q whether the interchange (3.13) is allowed or not. Studies with free field models suggest that there are no objections with commutators of vector and axial vector charges. In Chapter 4 we will connect the infinite-momentum sum rules to dispersion relations and we will see that the interchange of the limits in (3.13) corresponds to the absence of subtraction constants.

The study of the infinite-momentum limit has been started by Fubini and Furlan.[135] Its properties are further discussed in references (114) and (384).

3.3. Corrections to SU$_3$ Results

As pointed out in Section (3.1), the results of the SU$_3$ symmetry model are recovered from current algebras in the approximation of keeping only multiplet members as intermediate states in commutators. Higher states, in particular many-particle states, will provide the corrections. These could easily be evaluated if their weak interaction data were known; so far this is almost never the case.

Nevertheless some details can be learnt from the specific form of the correction term in (3.5); this term is a summation over states outside the considered multiplet:

$$\lim_{p \to \infty} \sum_{N'} \left[\frac{\langle a|Q^i|N'\rangle\langle N'|T^j|b\rangle}{(2\pi)^3 2E_{N'}} - \frac{\langle a|T^j|N'\rangle\langle N'|Q^i|b\rangle}{(2\pi)^3 2E_{N'}} \right]$$

$$= \sum_{N'} \left[\frac{\langle a|D^i|N'\rangle\langle N'|T^j|b\rangle}{i(m_a^2 - m_{N'}^2)} - \frac{\langle a|T^j|N'\rangle\langle N'|D^i|b\rangle}{i(m_{N'}^2 - m_b^2)} \right] \quad (3.16)$$

where we have rewritten

$$\langle a|j_0^i(0)|N'\rangle = \frac{\langle a|\partial^\mu j_\mu^i(0)|N'\rangle}{i(E_a - E_{N'})} \overset{\text{def}}{=} \frac{\langle a|D^i|N'\rangle}{i(E_a - E_{N'})} \quad (3.17)$$

and have taken the limit of infinite momentum with

$$2E_{N'}(E_a - E_{N'}) \to (m_a^2 - m_{N'}^2).$$

41

The current divergences are of first order in SU_3 violations; for a lowest-order estimate we may take SU_3-symmetry values for $\langle N'|T^j|b\rangle$ and $\langle a|T^j|N'\rangle$. It is impossible at present to consider all intermediate states in (3.16), and most authors take into account only two-particle states. In estimating the matrix element $\langle a|D|n_1 n_2\rangle$ two kinds of approximations have been used:

(a) Considering the symmetry-breaking only in external lines, as shown graphically:

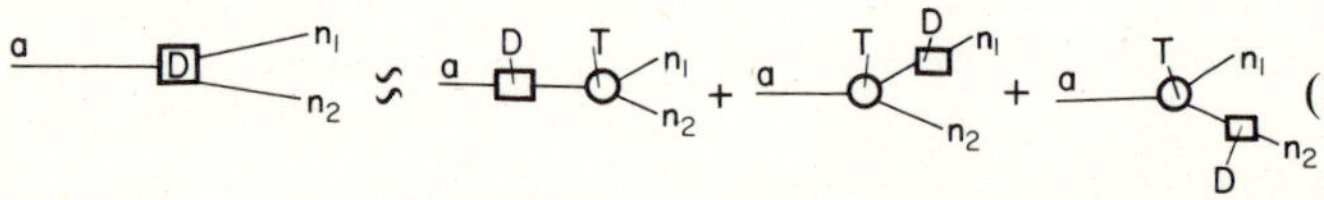

(3.18)

In (3.18) (T) denotes an SU_3-symmetric transition matrix element or coupling constant†; the idea of the approximation is that the mass shifts in external lines provide the dominant part of the SU_3 breaking.

One can also describe (3.18) in a spurion picture. A fictitious particle is associated with the operator D, then $\langle a|D|n_1 n_2\rangle$ represents the transition amplitude $n_1 + n_2 \rightarrow a + (D)$ and (3.18) approximates it by the nearest s,t and u poles.

(b) The spurion picture may be extended to use dynamical equations of Omnès type to estimate the couplings of (D) to higher states from its couplings within a multiplet, given by the current divergence in terms of the mass differences. Boiti and Rebbi[70] considered the case of baryons (B) and mesons (M): $B + M \rightarrow B + (D)$ and estimated from the baryon octet mass differences an effective coupling constant g^* for the spurion between the baryon octet and the $(\frac{3}{2})^+$ resonance decuplet. Knowing g^*, various SU_3 corrections can be estimated by keeping only the resonance decuplet in (3.16) and neglecting all other intermediate states.

Calculations of types (a) and (b) give rather small symmetry-breaking corrections, except for the Gell-Mann Okubo mass formula too small to be tested quantitatively. However, they can be used to show that the measured small value of the Cabibbo angle θ is not an effect of SU_3 violations, if current algebras are correct.

†Not to be confused with the tensor T^j.

So far the evidence that current algebras remain exact under SU$_3$ symmetry breaking is not yet completely convincing. Later we will find further confirmation by relating to each other different observed SU$_3$ violations with the help of supposedly exact current commutators.[319]

Finally we list the calculations so far performed for corrections of SU$_3$ and indicate the method of the authors in parentheses.

General methods for corrections to SU$_3$ results: (135), (149(a)), (319) (dispersion theory: Section 5.3).

Corrections to the Gell-Mann Okubo mass formulas: for baryons: (70(b)); (268) (model); for mesons: (149(a)).

Magnetic moments: 117(b).

Weak vector currents: for baryons: (92(b)), (260(a)); for mesons: (49(a)), (148(a)), (150) (dispersion theory: Section 6.2), (336) (dispersion theory: Section 6.2).

Weak axial vector currents: (91(b)); (92(b)); (101) (mixing effects: Chapter 7).

Vector meson decays: 321 (dispersion theory: Section 4.3).

Baryon Yukawa coupling constants (Low energy theorem): (74), (429); see also (424).

CHAPTER 4

The Dispersion Theory of Current Algebras, I

4.1. A Covariant Derivation of Sum Rules

In Chapter 3, when we derived direct sum rules by inserting intermediate states into the commutator of frame-dependent charges we found a dependence on the momentum of the external states. We will compare these sum rules with others derived in a manifestly covariant way. The essential step was taken by Fubini, Furlan and Rossetti[136] in constructing a dispersion theory of current algebras.

We consider two integrals

$$T_\mu = \int e^{iq_2 x} \vartheta(x_0) \langle p_2 | [j^i_\mu(x), h^j(0)] | p_1 \rangle (dx)^4, \qquad (4.1)$$

$$U = \int e^{iq_2 x} \vartheta(x_0) \langle p_2 | [D^i(x), h^j(0)] | p_1 \rangle (dx)^4, \qquad (4.2)$$

where $D^i(x)$ is the current divergence $\partial^\mu j^i_\mu(0)$ and $h^j(0)$ is some local operator; in applications it will be almost always a current or a current-divergence. For definiteness we will assume that it is a scalar (or pseudoscalar); then T_μ will be a vector (or axial vector) and U will be a scalar (or pseudoscalar), provided that the commutator is regular enough at the origin to allow a multiplication by the step-function $\vartheta(x_0)$. [For spacelike x the local commutator will, of course, vanish, but this is not yet sufficient for covariance in general[210]; here we will only deal with cases where (4.1) and (4.2) are covariant and will leave the more singular cases to Chapter 8.]

We perform a partial integration in (4.1) and drop the surface term at infinite positive time with the understanding that we give the vector $q_{2\mu}$ a small time-like imaginary part: $q_{2\mu} \to q_{2\mu} + i\epsilon_\mu$, $\epsilon^2 > 0$:

44

$$0 = iq_{2\mu}T_\mu + U + \int e^{iq_2 x}\delta(x_0)\langle p_2|[j_0^i(x),h^j(0)]|p_1\rangle (dx)^4. \quad (4.3)$$

The equal-time commutator comes from the differentiation of the step function $\vartheta(x_0)$. To form a charge in (4.3) we have to take the limit $q_{2\mu} \to 0$:

$$\lim_{q_{2\mu}\to 0} \int e^{iq_2 x}\delta(x_0)\langle p_2|[j_0^i(x),h^j(0)]|p_1\rangle = \langle p_2|[\int j_0^i(\bar{x},0)\,(d\bar{x})^3,h^j(0)]|p_1\rangle$$

$$= c^{ijk}\langle p_2|h^k(0)|p_1\rangle, \quad (4.4)$$

if we assume

$$[Q^i,h^j(0)] = c^{ijk}h^k(0). \quad (4.5)$$

Unless T_μ has a pole at $q_{2\mu} = 0$, the term $iq_{2\mu}T_\mu$ will vanish in (4.3). Such a pole arises only if there are intermediate states exactly degenerate in mass with the external ones. This can be easily seen by introducing intermediate states into T_μ and integrating (4.1). For simplicity we choose a frame where the time-like vector $q_{2\mu}$ has the form $(q_{20},\vec{0})$:

$$T_\mu = \sum_n \int_0^\infty dx_0\{e^{i(q_{20}+p_{20}-p_{n0})x_0}\langle p_2|j_\mu^i(0)|n\rangle\!\rangle\langle\!\langle n|h^j(0)|p_1\rangle\delta^3(p_2-p_n)$$

$$(2\pi)^3 - e^{i(q_{20}+p_{n0}-p_{10})x_0}\langle p_2|h^j(0)|n \quad n|j_\mu^i(0)|p_1\rangle\delta^3(p_n-p_1)(2\pi)^3\}$$

$$= \sum_n i\left\{\frac{\langle p_2|j_\mu^i(0)|n\rangle\!\rangle\langle\!\langle n|h^j(0)|p_1\rangle\delta^3(p_2-p_n)}{(q_{20}+p_{20}-p_{n0})}\right.$$

$$\left.\frac{\langle p_2|h^j(0)|n\rangle\!\rangle\langle\!\langle n|j_\mu^i(0)|p_1\rangle\!\rangle\delta^3(p_n-p_1)}{(q_{20}+p_{n0}-p_{10})}\right\}(2\pi)^3. \quad (4.6)$$

As $q_{20} \to 0$ the denominators in (4.6) take the form $(p_{20}-p_{n0})$ and $(p_{n0}-p_{10})$. They will vanish only if intermediate states contribute which are exactly degenerate in mass with the external ones. For the moment we exclude this case; we will return to it later. So we consider (4.3) in the form:

$$c^{ijk}\langle p_2|h^k(0)|p_1\rangle = -\lim_{q_{2\mu}\to 0} U$$

$$= -\lim_{q_{2\mu}\to 0}\int e^{iq_2 x}\vartheta(x_0)\langle p_2|[D^i(x),h^j(0)]|p_1\rangle(dx)^4. \quad (4.7)$$

The integral U in (4.7) is a relativistic scalar, so it can depend only on scalar variables. We have three independent vectors p_1,p_2,q_2, so we can form six scalars, two of which are the masses m_1^2 and m_2^2. To choose the remaining four scalar variables we introduce the vector q_1 by $p_1+q_1=p_2+q_2$ and take as independent scalars $\nu=\frac{1}{4}(p_1+p_2)(q_1+q_2)$; $t=(p_1-p_2)^2=(q_1-q_2)^2$; q_1^2; q_2^2. The limit $q_{2\mu}\to 0$ corresponds to $q_{1\mu}\to(p_2-p_1)_\mu$ and consequently $\nu\to\frac{1}{4}(m_2^2-m_1^2)$; $q_1^2\to t$, $q_2^2\to 0$.

The integral $U(\nu,t,q_1^2,q_2^2)$ is the Fourier transform of a retarded commutator. The analyticity properties of such integrals in the variables ν and t have been extensively investigated, because in an LSZ representation (see Appendix) scattering amplitudes can be written as Fourier transforms over retarded commutators of Heisenberg field operators. Rigorous analyticity proofs are available for very simple kinematic configurations, for instance for the forward scattering of massless particles ($q_1^2=q_2^2=0$) at

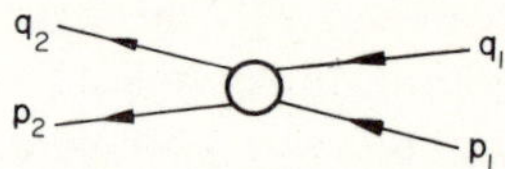

spinless targets: $p_1=p_2$. The complexity of the arguments increases considerably if more complicated configurations are considered.

For analyticity proofs we refer to the literature ($S\,1-S\,3$ and references given there); here we will only state the results. For scattering amplitudes it is natural to fix the masses q_1^2 and q_2^2; t is also fixed in general and dispersion relations in ν are derived. It can be shown (rigorously or heuristically — depending on the complexity of the kinematics) that retarded commutators like $U(\nu,t,q_1^2,q_2^2)$ are analytic in the complex ν-plane apart from a cut along the real ν-axis. If all particles in the problem are spinless then the ν-discontinuity of the retarded commutator is given by a related integral

$$u(\nu,t,q_1^2,q_2^2)=\int e^{iq_2x}\langle p_2|[D^i(x),h^j(0)]|p_1\rangle\,(dx)^4. \qquad (4.8)$$

We may then write a dispersion relation

$$U(\nu,t,q_1^2,q_2^2)=\frac{1}{2\pi i}\int\frac{u(\nu',t,q_1^2,q_2^2)d\nu'}{\nu'-\nu}$$

provided the integral converges and provided $U(\nu,t,q_1^2,q_2^2)$ tends to zero at large ν. Otherwise we need subtraction polynomials. In the applications to current algebras we will assume that unsubtracted dispersions relations hold; then we can represent the commutator (4.7) as

$$c^{ijk}\langle p_2|h^k(0)|p_1\rangle = -U(\tfrac{1}{4}\Delta m^2,t,t,0) = \frac{i}{2\pi}\int \frac{u(\nu',t,t,0)d\nu'}{(\nu'-\tfrac{1}{4}\Delta m^2)}. \tag{4.10}$$

If the dispersion relations have to be subtracted then they would be less useful for current algebras because the information contained in the commutator would then go into the subtraction constant and we would not obtain a sum rule.

If the external particles have spin or if $h^j(0)$ is not a scalar, as required initially, most of the arguments leading to dispersion relations of the type (4.10) remain unchanged. The main difference is that $U(\nu,t,q_1^2,q_2^2)$ in (4.7) and $u(\nu,t,q_1^2,q^2)$ in (4.8) will now contain several kinematic factors; for instance with $h^j(0)=j_\mu^j(0)$ we will have (keeping $\langle p_2|$ and $|p_1\rangle$ spinless):

$$U_\mu(\nu,t,q_1^2,q_2^2) = \int \vartheta(x_0)e^{iq_2x}\langle p_2|\,[D^i(x),j_\mu^j(0)]\,|p_1\rangle\,(dx)^4$$

$$= U^{(1)}(\nu,t,q_1^2,q_2^2)p_{1\mu} + U^{(2)}(\nu,t,q_1^2,q_2^2)p_{2\mu}$$

$$+ U^{(3)}(\nu,t,q_1^2,q_2^2)\,q_{2\mu} \tag{4.11}$$

$u_\mu(\nu,t,q_1^2,q_2^2)$ has a similar decomposition and so has the current matrix-element $\langle p_2|j_\mu^k(0)|p_1\rangle$ in (4.7)

$$u_\mu(\nu,t,q_1^2,q_2^2) = u^{(1)}(\nu,t,q_1^2,q_2^2)p_{1\mu} + u^{(2)}(\nu,t,q_1^2,q_2^2)p_{2\mu}$$

$$+ u^{(3)}(\nu,t,q_1^2,q_2^2)\,q_{2\mu} \tag{4.12}$$

$$\langle p_2|j_\mu^k(0)|p_1\rangle = F^{(1)}(t)p_{1\mu} + F^{(2)}(t)p_{2\mu} \tag{4.12A}$$

In dispersion relations the invariants of equal kinematic forms are usually connected to each other. So we have two relations instead of (4.10):

$$c^{ijk}F^{(1)}(t) = -U^{(1)}(\tfrac{1}{4}\Delta m^2,t,t,0) = \frac{i}{2\pi}\int \frac{u^{(1)}(\nu',t,t,0)}{(\nu'-\tfrac{1}{4}\Delta m^2)}d\nu' \tag{4.13}$$

$$c^{ijk}F^{(2)}(t) = -U^{(2)}(\tfrac{1}{4}\Delta m^2,t,t,0) = \frac{i}{2\pi}\int \frac{u^{(2)}(\nu',t,t,0)}{(\nu'-\tfrac{1}{4}\Delta m^2)}d\nu'. \tag{4.14}$$

47

A third relation:

$$-U^{(3)}(\tfrac{1}{4}\Delta m^2,t,t,0) = \frac{i}{2\pi}\int\frac{u^{(3)}(v',t,t,0)}{(v'-\tfrac{1}{4}\Delta m^2)}dv'. \qquad (4.15)$$

is not of direct use for current algebras, since in (4.7) the equal time commutator is related to the limit $q_{2\mu}\to 0$, where the third kinematic factor vanishes, so there is no prediction made for $U^{(3)}(\tfrac{1}{4}\Delta m^2,t,t,0)$.

The choice of kinematic forms in (4.11) is not quite arbitrary. First, we have to avoid kinematical singularities in the invariant matrix elements; methods for this have been developed.†
Second, we should choose a decomposition where we have as many rapidly converging dispersion relations as possible. The criteria for such a choice are still under investigation; what is known will be presented in Section 9.5.

The prescriptions of the dispersion theory are now almost complete; the only gap is the case of mass-degeneracy between external and intermediate particles, explicitly excluded at the beginning of this section. In this case we have two undefined terms: (a.) $\lim q_{2\mu}T_\mu$ in (4.3) and (b.) a term with vanishing numerator and denominator in the dispersion integral (4.10). Depending on the details of the limiting process one or the other term may be taken to vanish. There are different consistent prescriptions (leading to the same results of course); we recommend to avoid these rather tricky problems by using electromagnetic mass differences or by introducing artificially fictitious small mass differences δm between external and intermediate states and to apply the analysis as described above (with $q_{2\mu}T_\mu \to 0$); the result will always be independent of δm, which may be taken to vanish in the end of the calculation. In proposing this prescription we use a principle of continuity in physics between exact and weakly broken symmetries. One can easily convince oneself with examples that this prescription gives correct results; it will be justified more generally in Section 9.1.

For conserved charges Q^i group invariance allows us to deduce the consequences of the commutator $[Q^i,h^j(0)] = c^{ijk}h^k(0)$ directly by the Wigner Eckart theorem. The dispersion relations of this section are designed to deal with non-conserved charges. They will also work for conserved charges if one proceeds consis-

---

†Using, for instance, the helicity formalism.[S 5]

tently (without obtaining anything new of course). We may either insist on current conservation $D^i = 0$, then U vanishes, and we have to evaluate the poles of T_μ; or alternatively we may introduce a small symmetry breaking, then $q_{2\mu}T_\mu$ goes to zero with non-degenerate masses and we get the same results from (4.10), if we let the artificial symmetry breaking go to zero in the end. Both ways are rather clumsy compared with the direct application of the Wigner–Eckart theorem.

In this section it was unavoidable to consider special cases first for the demonstration of the essential ideas in the dispersion theory of current algebras and then to generalize them. It might be useful to remember which assumptions remain necessary:

(a) covariance of T_μ and U, implying a certain regularity of the local commutator (page 44). This question will be further investigated in Chapter 8.

(b) unsubtracted dispersion relations in ν for suitably chosen invariants in U (pp. 46/48). This will remain essentially an assumption; some further comments will be made in Sections 8.2 and 9.5.

(c) a prescription to deal with degenerate intermediate states; this prescription will be justified in Section 9.1.

As already mentioned, this dispersion theory of current algebras has been proposed by Fubini, Furlan and Rossetti;[136–137] for some further remarks† see (18) and (233). Schroer and Stichel have investigated the theory using the Jost–Lehmann–Dyson representation.[350]

Applications will be given in the following sections of this chapter and in Chapter 5 so that there will be sufficient opportunity to go through the technicalities of the method.

4.2. The Relationship between Direct Sum Rules and Dispersion Sum Rules

There is an immediate connexion between the dispersion sum rules of Section 4.1 and direct sum rules taken at infinite momentum of the external particles. For the simple case of spinless particles $\langle p_2| \ldots |p_1 \rangle$ and a scalar operator $h^j(0)$ we will explicitly demonstrate their equivalence. We write a direct sum rule from (4.5)

$$[Q^i, \int e^{-iq_1 x}h^j(x)] = c^{ijk} \int e^{-iq_1 x}h^k(x)\,(dx)^3. \qquad (4.16)$$

†Recent contribution: (417).

$$\sum_n {}^{\cdot} \left\{ \frac{\langle p_2|j_0^i(0)|n\rangle\langle n|h^j(0)|p_1\rangle \delta^3(p_n-p_2)\delta^3(p_n-p_1-q_1)}{2E_n} \right\}$$

$$-\sum_n {}^{\cdot} \left\{ \frac{\langle p_2|h^j(0)|n\rangle\langle n|j_0^i(0)|p_1\rangle \delta^3(p_2-p_n-q_1)\delta^3(p_n-p_1)}{2E_n} \right\}$$

$$= c^{ijk}\langle p_2|h^k(0)|p_1\rangle \delta^3(p_1+q_1-p_2) \qquad (4.17)$$

We introduce current divergences:

$$\langle p_2|j_0^i(0)|n\rangle = \frac{\langle p_2|D^i(0)|n\rangle}{i(E_2-E_n)}$$

(with $\bar{p}_2 = \bar{p}_n$) and pass to the limit of infinite momentum in the z-direction. We choose q_1 in the x-direction, then at infinite momentum the energy differences between the states tend to zero and we will have momentum transfers $t = q_1^2$ in the matrix elements $\langle p_2|h^k(0)|p_1\rangle$ and $\langle n|h^j(0)|p_1\rangle$, $\langle p_2|h^j(0)|n\rangle$; the momentum transfers in $\langle p_2|D^i(0)|n\rangle$ and $\langle n|D^i(0)|p_1\rangle$ tend to zero. The denominators $2E_n$ and (E_2-E_n) will combine in the limit $2E_n \cdot (E_2 - E_n) \to (m_2^2 - m_n^2)$. In the sum over states we first integrate over internal degrees of freedom; introducing

$$\sum_{\text{int}} \delta(m_n-W)\langle p_2|D^i(0)|n\rangle_{\Delta^2=0}\langle n|h^j(0)|p_1\rangle_{\Delta^2=t} \overset{\text{def}}{=} H_W(D^i,h^j);$$

$$(4.18)$$

the momentum integration is absorbed by the δ-functions and mass integration remains to be done:

$$c^{ijk}\langle p_2|h^k(0)|p_1\rangle = \frac{1}{i}\int dW\left(\frac{H_W(D^i,h^j)}{m_2^2-W^2} - \frac{H_W(h^j,D^i)}{W^2-m_1^2}\right). \quad (4.19)$$

All quantities appearing in (4.19) are relativistic scalars, so the equation is independent of the state of motion of $\langle p_2|$ and $|p_1\rangle$. We will show its equivalence to (4.10). We integrate $u(\nu',t,q_1^2,q_2^2)$ by introducing intermediate states

$$u(\nu', t, q_1^2, q_2^2) =$$

$$(2\pi)^4 \sum_n \left\{ \frac{\delta^4(p_2+q_2-p_n)\langle p_2|D^i(0)|n\rangle\langle n|h^j(0)|p_1\rangle}{(2\pi)^3 2E_n} \right.$$

$$\left. -\frac{\delta^4(p_n+q_2-p_1)\langle p_2|h^j(0)|n\rangle\langle n|D^i(0)|p_1\rangle}{(2\pi)^3 2E_n} \right\}.$$

$$(4.20)$$

In (4.20) we indicated that the momentum q_2 is associated with D^i and q_1 with h^j. Consequently there will be momentum transfers $q_1^2 = t$ in the matrix elements of t^j and $q_2^2 = 0$ in the matrix elements of D', as in (4.18) and (4.19). Again we split the intermediate state summation into an integration over internal degrees of freedom, total mass W and spatial momentum (absorbed by the spatial δ-function). We obtain

$$u(\nu',t,t,0) = 2\pi \int dW \left(\frac{H_W(D^i,h^j)}{2E_W} \delta(E_W - p_{20} - q_{20}) \right.$$

$$\left. - \frac{H_W(h^j,D^i)}{2E_W} \delta(E_W + q_{20} - p_{10}) \right) \quad (4.21)$$

E_W is the energy of the intermediate state $|n\rangle$ of mass W fixed by $q_1^2 = t$ and $q_2^2 = 0$. We substitute (4.21) into (4.10). We have introduced $\nu' = \frac{1}{4}(p_1 + p_2)(q_1 + q_2)$ and have fixed $t = (p_2 - p_1)^2 = (p_2 - p_1)(q_1 - q_2)$. We can combine these expressions to

$$2(p_1q_1 + p_2q_2) = 4\nu' - t; \quad 2(p_1q_2 + p_2q_1) = 4\nu' + t. \quad (4.22)$$

From the δ- functions in (4.20) we obtain the conditions

$$m_n^2 = m_1^2 + 2p_1q_1 + q_1^2; \quad m_n^2 = m_1^2 - 2p_1q_2 + q_2^2;$$
$$m_n^2 = m_2^2 + 2p_2q_2 + q_2^2; \quad m_n^2 = m_2^2 - 2p_2q_1 + q_1^2 \quad (4.23)$$

for contributions to respectively the first or second part of the commutator. Substituting the values $q_2^2 = 0, q_1^2 = t$ we can introduce (4.23) into (4.22):

$$2m_n^2 - m_1^2 - m_2^2 - t = 4\nu' - t; \quad m_1^2 + m_2^2 - 2m_n^2 + t = 4\nu' + t. \quad (4.24)$$

This determines the denominators in (4.10).

$$\nu' - \tfrac{1}{4}(m_2^2 - m_1^2) = \tfrac{1}{2}(m_n^2 - m_2^2); \quad \nu' - \tfrac{1}{4}(m_2^2 - m_1^2) = \tfrac{1}{2}(m_1^2 - m_n^2) \quad (4.25)$$

in the first and second part of the commutator, respectively. Finally we have to evaluate the integrals

$$\int \frac{dv'}{2E_W}\delta(E_W-p_{20}-q_{20}) \quad \text{and} \quad \int \frac{dv'}{2E_W}\delta(E_W-p_{10}+q_{20})$$

$$=\int \frac{dv'}{2E_W}\delta(E_W-p_{20}+q_{10}). \quad (4.26)$$

Equation (4.21) is a covariant expression, therefore we may choose a special system and we take the rest system of $\langle p_2|$. By $(p_1+q_1)^2=(p_2+q_2)^2$ we express v' in terms of p_2q_2,

$$p_2q_2=v'+\tfrac{1}{4}(m_1^2+q_1^2-m_2^2-q_2^2-t)=v'+\tfrac{1}{4}(m_1^2-m_2^2), \quad (4.27)$$

so we have

$$q_{20}=(v'+\tfrac{1}{4}(m_1^2-m_2^2))/m_2$$

$$E_W=\sqrt{(m_n^2+\bar{p}_n^2)}=\sqrt{(m_n^2+\bar{q}_2^2)}=\sqrt{(m_n^2+q_{20}^2-q_2^2)}$$
$$\text{(from the first part of eq. (4.20)),}$$

and finally

$$\frac{\partial}{\partial v}(E_W-m_2-q_{20})=\frac{q_{20}}{m_2E_W}-\frac{1}{m_2}=\frac{1}{m_2E_W}(q_{20}-E_W). \quad (4.28)$$

So we obtain:

$$\int \frac{dv'}{2E_W}\delta(E_W-p_{20}-q_{20})=\tfrac{1}{2}; \quad (4.29)$$

the second integral in (4.26) gives the same result. Now we substitute equations (4.21), (4.25) and (4.29) into (4.10) and get (4.19):

$$c^{ijk}\langle p_2|h^k(0)|p_1\rangle=i\int dW\left(\frac{H_W(D^i,h^j)}{(W^2-m_2^2)}-\frac{H_W(h^j,D^i)}{(m_1^2-W^2)}\right). \quad (4.30)$$

Direct sum rules taken at finite momenta have variable momentum transfers in the matrix elements of the currents, dependent on the mass of the intermediate states. For a static sum rule $(\bar{p}=0)$ we have $\Delta^2=(m_{1,2}-m_n)^2$ because intermediate states are taken at rest. To reproduce these contributions in a dispersion relation we can no longer fix the "masses" q_1^2 and q_2^2; we lose the analogy to dispersion relations of scattering matrix elements.

Nevertheless it is still possible to reproduce direct sum rules taken at arbitrary momenta $\bar{p}$ in the form of covariant dispersion integrals; we have just to disperse along a different surface in the space of the variables ν,t,q_1^2,q_2^2. To demonstrate the essential ideas we may choose the simple configuration $p_1 = p_2$, $q_1 = q_2$: $\nu = p \cdot q, t = 0, q_1^2 = q_2^2 = q^2$; $u(\nu,t,q_1^2,q_2^2)$ will be denoted by $u(\nu,q^2)$.

If we have a static sum rule we want to reproduce contributions of intermediate states at rest, i.e. $\bar{q} = 0$. This specifies the surface $q^2 = q_0^2 = (\nu/m_p)^2$; the dispersion relation would read:

$$c^{ijk}\langle p|h^k(0)|p\rangle = -U(0,0) = \frac{i}{2\pi} \int \frac{u(\nu',(\nu'/m_p)^2)}{\nu'}d\nu'. \qquad (4.31)$$

For a general momentum $\bar{p}$ we have to take the surface:

$$q^2 = [\sqrt{(m_p^2 + 2\nu + q^2 + \bar{p}^2)} - \sqrt{(m_p^2 + \bar{p}^2)}]^2. \qquad (4.32)$$

Derivation. In (4.20) q^2 is the momentum transfer between the intermediate and external states

$$q^2 = [\sqrt{(m_n^2 + \bar{p}^2)} - \sqrt{(m_p^2 + \bar{p}^2)}]^2 \qquad (\alpha).$$

m_n^2 has to be eliminated from (α) by squaring $p_n = p + q$. This leads to (4.32).

So all direct sum rules can be derived covariantly; this is why we do not adopt the terms "frame-dependent" or "non-covariant" sum rules which can sometimes be found in the literature. If the analyticity properties of $u(\nu,q^2)$ are suitable then the surface (4.32) can be shifted in the space of (ν,q^2), and different direct sum rules will not provide independent tests of current algebras.

For practical purposes only the sum rules with fixed momentum transfers q_1^2 and q_2^2 in the currents are useful. As we will see in Section 5.1, we can represent the matrix elements of currents and divergences by their single-particle-poles, provided the momentum transfer is small. Singularities at high momentum transfer in the current matrix elements are determined by many-particle reactions and cannot be estimated at present.

Some complications arise in the comparison of direct sum rules and dispersion sum rules, if the external particles have spin, or more generally if there is more than one independent kinematic form for dispersion relations. There are dispersion sum rules which cannot be obtained as direct sum rules (a) and direct sum rules which are not obtained by the usual dispersion methods of Section 4.1. We do not want to emphasize these discrepancies, it may well be that one or the other method has not yet reached its most efficient form; we just illustrate them to make the reader familiar with the technicalities of applications.

(a) *The Derivation of the Gell-Mann Okubo Mass Formula for Baryons*

In Section 3.2 we have derived the Gell-Mann Okubo mass formula from the vanishing commutator (3.9)

$$\left[Q^{K+},\frac{dQ^{K+}}{dt}\right] = [Q^{K+},\ \int (dx)^3 D^{K+}(x)] = 0. \qquad (4.33)$$

For dispersion relations we need a local commutator, so we postulate:

$$[Q^{K+}, D^{K+}(0)] = 0. \qquad (4.34)$$

We take (4.34) between a proton and a Ξ^--hyperon:

$$U(\nu,t,q_1^2,q_2^2) = \int \vartheta(x_0) e^{iq_2 x}\langle P(p_2)|[D^{K+}(x),D^{K+}(0)]|\Xi^-(p_1)\rangle (dx)^4$$

$$= \bar{u}_P(p_2)(U^{(1)}(\nu,t,q_1^2,q_2^2) + \gamma q_2 U^{(2)}(\nu,t,q_1^2,q_2^2))u_\Xi(p_1).$$

$$(4.35)$$

Current algebras (4.34) give the limit of (4.35) as $q_{2\mu} \to 0$

$$0 = U^{(1)}(\tfrac{1}{4}(m_P^2 - m_\Xi^2),t,t,0). \qquad (4.36)$$

The second kinematic form vanishes and therefore no statement is made about $U^{(2)}(\tfrac{1}{4}(m_P^2 - m_\Xi^2),t,t,0)$. We choose $t = 0$ and according to the prescriptions of Section 4.1 we separate $u(\nu',t,t,0)$ into the same two kinematic forms (4.35):

54

$$u(\nu,0,0,0)$$

$$= (2\pi)^4 \sum_n \{\langle P(p_2)|D^{K+}(0)|n\rangle\langle n|D^{K+}(0)|\Xi^-(p_1)\rangle \delta^4(p_2+q_2-p_n)$$

$$- \langle P(p_2)|D^{K+}(0)|n\rangle\langle nD^{K+}(0)|\Xi^-(p_1)\rangle\delta^4(p_n+q_2-p_1)\}. \quad (4.37)$$

Contributions will come from Λ and Σ° in the baryon octet and from higher states. We calculate in detail the Λ-contribution to the first part of (4.37). For the vector current

$$\langle P(p_2)|j_\mu^{K+}(0)|\Lambda(p_n)\rangle$$

we use the notation of (1.7):

$$\langle P(p_2)|j_\mu^{K+}(0)|\Lambda(p_n)\rangle = \bar{u}_P(p_2)(F_{1,P\Lambda}(q_2^2)\gamma_\mu + F_{2,P\Lambda}(q_2^2)\sigma_{\mu\nu}(p^\Lambda-p^p)_\nu$$

$$+ F_{3,P\Lambda}(q_2^2)(p^\Lambda-p^p)_\mu)u_\Lambda(p_n). \quad (4.38)$$

At zero momentum transfer only the first term contributes to the divergence

$$\langle P(p_2)|D^{K+}(0)|\Lambda(p_n)\rangle = i(m_P-m_\Lambda)\bar{u}_P(p_2)F_{1,P\Lambda}(0)u_\Lambda(p_n).$$
$$(4.39)$$

We use the Ademollo–Gatto theorem to set $F_{1,P\Lambda}(0) = \sqrt{\tfrac{3}{2}}$; because (4.39) contains the factor $(m_P - m_\Lambda)$ we are neglecting a third-order quantity. Summing over the Λ-polarization we find the contribution to the first term in (4.37):

$$-(2\pi)(m_P-m_\Lambda)(m_\Lambda-m_\Xi)\cdot\tfrac{3}{2}\bar{u}_P(p_2)\frac{(\gamma p_\Lambda+m_\Lambda)}{2E_\Lambda}u_\Xi(p_1)\cdot$$
$$\delta(p_{20}+q_{20}-p_{\Lambda 0}). \quad (4.40)$$

In (4.40) we separate $\gamma p_1 = \gamma p_2 + \gamma q$ and project the contribution to $u^{(1)}$:

$$-(2\pi)(m_P^2-m_\Lambda^2)(m_\Lambda-m_\Xi)\cdot\tfrac{3}{2}\frac{\delta(p_{20}+q_{20}-p_{\Lambda 0})}{2E_\Lambda}+0(\lambda^3). \quad (4.41)$$

Substituting (4.41) into a dispersion integral of the type (4.13) and remembering the results (4.25) and (4.29) (which are independent of spins and kinematic forms) we find the Λ-contribution to the dispersion integral from (4.41):

$$i\tfrac{3}{2}(m_\Lambda - m_\Xi) + 0(\lambda^3).\qquad(4.42)$$

In the same way we can calculate the Σ^0 contribution to the first term of (4.20); it comes out: $i\tfrac{1}{2}(m_\Sigma - m_\Xi) + 0(\lambda^3)$. Combining this with the second term, we get linear mass formulas:

$$0 = i\,(3m_\Lambda + m_\Sigma - 2m_P - 2m_\Xi) + 0(\lambda^3)$$
$$+ \text{(continuum contributions).}\qquad(4.43)$$

This result is clearly different from the direct sum rule at infinite momentum, which gives a square mass formula (see (3.11)).

The source of this discrepancy is the different treatment of the term proportional to γq in (4.40). In the dispersion method we have separated it off as contributing to $U^{(2)}(\nu, t, q_1^2, q_2^2)$. In direct sum rules there is no separation of kinematic forms. The term $\bar{u}_P(\bar{p})(\gamma p_\Lambda - \gamma p_P)u_\Xi(\bar{p})$, which corresponds to the separated term in (4.40), has the form

$$\bar{u}_P(\bar{p})\gamma_0(E_P - E_\Lambda)u_\Xi(\bar{p}) \approx \frac{1}{2\bar{p}}(m_P^2 - m_\Lambda^2)\bar{u}_P(\bar{p})\gamma_0 u_\Xi(\bar{p}).\qquad(4.44)$$

In the limit $\bar{p} \to \infty$ (4.44) will not vanish because $u_P(\bar{p})\gamma_0 u_\Xi(\bar{p})$ diverges like $\bar{p}$. Inclusion of (4.44) then leads to the square mass formula.

Instead of the form factors in (4.35) one may of course choose others and reproduce the square mass formula from dispersion theory. With mesons there is a unique result: the square mass formula; with baryons the ambiguity of the kinematic forms allows the deduction of a linear mass formula (with different continuum contributions of course to balance the difference). This ambiguity has been discovered by Fubini, Furlan and Rossetti[136] (though these authors do not call it an ambiguity).

We want to give yet another example of the incomplete matching of the two methods in their present form.

(b) *Sum Rules for Vector Mesons*

Vector mesons have three helicity states. Charge-commutators and commutators between charges and current time-components are invariant against rotations; this means that in the direction of the common momentum external and intermediate states must have the same helicity. Sum rules obtained from external vector meson states with helicities (± 1) are related to each other, the sum rule obtained from helicity 0 states is different; only for this one do we have contributions from spinless intermediate states.

If on the other hand we form a charge-current commutator in dispersion theory

$$U_\mu(\nu, q^2) = \int \vartheta(x_0)e^{iqx}\langle V'(p)|[D^i(x), j^i_\mu(0)]|V(p)\rangle(dx)^4$$

$$= (\epsilon \cdot \epsilon')p_\mu A(\nu, q^2) + \{(\epsilon'q)\epsilon_\mu$$

$$+ (\epsilon q)\epsilon'_\mu\} B(\nu, q^2) + \cdots \qquad (4.45)$$

we have only one vector form: $(\epsilon \cdot \epsilon')p_\mu$ which is bilinear in the polarization vectors ϵ and ϵ' and does not vanish at $q \to 0$; so unless we develop the prescriptions of the dispersion theory further we get only one sum rule (corresponding to transverse polarizations). This phenomenon has been independently discovered and discussed by Gasiorowicz and Geffen[156] and by Taylor.[372] Both types of sum rules give acceptable results when saturated with resonances;[130] but with the present experimental knowledge we cannot regard them as confirmed.

4.3. Sum Rules for Vertex Functions

In this section we discuss the modifications which are necessary when considering commutators between particle-states and the vacuum. Due to momentum conservation, only charge-current commutators can be used:

$$\langle p|[Q^i(0), h^j(0)]|0\rangle = c^{ijk}\langle p|h^k(0)|0\rangle$$

$$= \sum_n \{\langle p|j^i_0(0)|n\rangle\frac{\delta^3(p_n - p_p)}{2E_n}\langle n|h^j(0)|0\rangle$$

$$- \langle p|h^j(0)|n\rangle\frac{\delta^3(p_n)}{2m_n}\langle n|j^i_0(0)|0\rangle\}. \qquad (4.46)$$

Going to the limit $p \to \infty$ we find zero momentum transfer in $\langle p|j^i_0(0)|n\rangle$, but infinite momentum transfer in $\langle p|h^j(0)|n\rangle$:
$\Delta^2 = (E_p - m_n)^2 - \bar{p}^2 = m_p^2 - 2E_pm_n + m_n^2 \to \infty$. In many cases one can assume that the matrix element $\langle p|h^j(0)|n\rangle$ will vanish at infinite momentum transfer, so only the first part of the commutator will contribute.

Equivalently for dispersion sum rules we have to consider:

$$U(\nu,q^2) = \int \vartheta(x_0)e^{iqx}\langle p|[D^i(x),h^j(0)]|0\rangle (dx)^4. \quad (4.47)$$

There are again two relativistic scalars in the problem, $\nu = p \cdot q$ and q^2. We disperse in ν with $q^2 = 0$:

$$u(0,0) = \frac{1}{2\pi i}\int \frac{u(\nu',0)}{\nu'}d\nu' = -c^{ijk}\langle p|h^k(0)|0\rangle \quad (4.48)$$

with

$$u(\nu',0) = \int e^{iq'x}\langle p|[D^i(x),h^j(0)]|0\rangle (dx)^4$$
$$= (2\pi)^4 \sum \{\langle p|D^i(0)|n\rangle\langle n|h^j(0)|0\rangle\delta^4(p+q'-p_n)$$
$$- \langle p|h^j(0)|n\rangle\langle n|D^i(0)|0\rangle\delta^4(p_n'+q')\}. \quad (4.49)$$

Due to $q'^2 = 0$ only zero mass intermediate states would contribute to the second part of the commutator, so it vanishes as in (4.46). A considerable simplification occurs in the first part, both in (4.46) and in (4.49). Depending on the rotational properties of $h^j(0)$, only states of specific low spins can contribute; a vector, for instance, can connect the vacuum only to $J^P = 0^+$ and 1^- states.

The justification of the dispersion relations (4.48) can no longer be supplied by the analogy to a scattering matrix element, but in an LSZ representation (4.47) has the form of a vertex matrix element $\langle p|h^j(0)|{}``D(q^2)"\rangle$ of $h^j(0)$ between $\langle p|$ and a virtual state $|``D"\rangle$ of mass q^2. The momentum transfer to be dispersed in would be $\Delta^2 = (p-q)^2 = m_p{}^2 - 2\nu + q^2$.

For applications see references: (150), (163), (320–2).

CHAPTER 5

Current Algebra Sum Rules with PCAC

5.1. The Principle of PCAC or Pion Pole Dominance

So far we have developed current algebras more or less
theoretically, apart from discussing their relation to the SU_3
symmetry model. We have not yet given any convincing tests for
the correctness of the theory. Current commutators give rise
to an immense number of sum rules connecting matrix elements
of weak or electromagnetic reactions (as in Section 2.4), but the
relevant matrix elements of many-particle intermediate states
are almost never known. We must therefore combine current
algebras with additional dynamical information in order to
approximate the sum rules in a testable form.

We will represent matrix elements of a current $\langle a|j_\mu|b\rangle$ or a
divergence $\langle a|D|b\rangle$ by dispersion relations in suitable relativistic
invariants. Now for single-particle states it can be shown that
the form factors $H(\Delta^2)$ of $\langle a|j_\mu|b\rangle$ and $\langle a|D|b\rangle$ are analytic
functions of the momentum transfer $\Delta^2 = (p_a - p_b)^2$ with singularities only along the real axis (for $\Delta^2 \geqslant \Delta_0^2$). Assuming suitable
convergence properties, the form factors can be represented by
their discontinuities over the real axis

$$H(\Delta^2) = \frac{1}{2\pi i} \int\limits^{\infty} \frac{\operatorname{disc} H(\Delta'^2)}{\Delta'^2 - \Delta^2} d\Delta'^2. \tag{5.1}$$

The origins of such singularities lie in different dynamical
mechanisms. (Again we will give mainly results and refer to the
literature: S 1–S6 for justification.) The practically most important singularities are given by normal thresholds: states with
the quantum numbers of the channel $(a\bar{b})$,† which provide real

---

†$\bar{b}$ denotes the antiparticle of b. The channel $(\bar{a}b)$ gives equivalent results by
charge-conjugation and time-reversal invariance of strong interactions.

or virtual intermediate steps of the chain: $\langle a\bar{b}|T^{+}|n\rangle\langle n|j_{\mu}|0\rangle$. Crossing principles imply that the momentum transfer Δ^2 in $H(\Delta^2)$ can be analytically continued to the invariant energy $(p^a+p^b)^2$ in the channel $(a\bar{b})$. Single-particle intermediate states of mass m with the quantum numbers of the current j_{μ} produce poles in $H(\Delta^2)$ at $\Delta^2 = m^2$, while continua produce cuts which are often approximated by poles at the positions of resonances. The residues of the poles factorize into $\langle a|t|bm\rangle\langle m|j|0\rangle$, where $\langle a|t|bm\rangle$ is a reduced transition matrix element (not containing the non-covariant normalization factors) or an effective coupling constant if the transition is virtual. This interpretation of the pole residue is easily understandable in perturbation theory. It just corresponds to the graphs

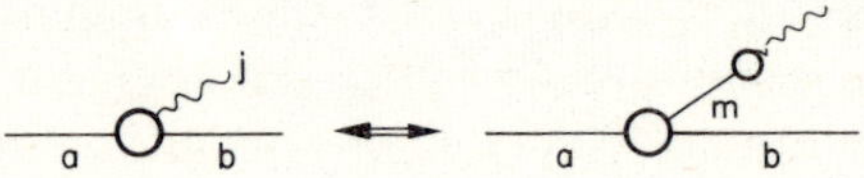

$$\tag{5.2}$$

but it can also be justified from more fundamental principles.

There are other singularities which are not generated by the above mechanism. For example, there are anomalous thresholds which can arise whenever $|a\rangle$ and $|b\rangle$ are many-particle states. Quantitatively they do not seem to be very important; in our applications we will rarely reach the level of accuracy at which they would be significant.

It should now be clear why we prefer to work with current matrix elements at low momentum transfer. The singularities nearest to the origin in (5.1) are the poles produced by the low-mass single-particle states. Their residues contain a decay constant $\langle m|j|0\rangle$ and a strong-interaction transition matrix element $\langle a|t|bm\rangle$; in some cases both of these are known. The singularities at high momentum transfers, created by many-particle states, cannot be expressed by measured quantities. For the neighbourhood of the origin in (5.1) we assume that the known contributions from nearby poles dominate the un-known ones from distant cuts. In practice this approximation is quite successful.

A particularly favourable case is provided by the matrix elements of the axial current divergence with π-meson quantum numbers: $\bar{D}^i$ $(i=1,2,3)$. They have a pole in momentum transfer at the π-meson mass: $\Delta^2 = m_\pi^2$ and a cut starting at $9m_\pi^2$, if anomalous thresholds are ignored.

$$\langle a|\bar{D}^i|b\rangle = -\frac{\langle a|t|b\pi\rangle\langle\pi|\bar{D}^i|0\rangle}{m_\pi^2-\Delta^2} + \frac{1}{2\pi i}\int\limits_{(3m_\pi)^2}^{\infty}\frac{f(\Delta'^2)}{\Delta'^2-\Delta^2}\,d\Delta'^2. \quad (5.3)$$

The matrix element $\langle\pi|\bar{D}^i|0\rangle$ is determined by the pion decay constant f_π for which we introduce the notation†

$$\langle 0|\bar{j}_\mu^1 - i\bar{j}_\mu^2|\pi^+\rangle = if_\pi p_\mu^\pi. \qquad (5.4)$$

From the origin the distances to pole and cut are proportional to 1:9. Although this is certainly not a sufficient argument to conclude that the cut contribution will be about 10 per cent of the pole contribution, such estimates seem to work in practice. As an example we derive the Goldberger–Treiman formula from the assumption of pion pole dominance in nucleon form factors. With the axial vector current between nucleons in the form of (1.7)

$$\cos\vartheta\langle P(p^f)|(\bar{j}_\mu^1+i\bar{j}_\mu^2)|N(p^i)\rangle$$
$$= \bar{u}_P(p^f)\,(G_{1,PN}^W(\Delta^2)\gamma^5\gamma_\mu + G_{2,PN}^W(\Delta^2)\,\gamma^5\sigma_{\mu\nu}(p^i-p^f)_\nu$$
$$+ G_{3,PN}^W(\Delta^2)\gamma^5(p^i-p^f)_\mu)\,u_N(p^i)$$

We obtain for the divergence

$$i\cos\vartheta\langle P(p^f)|(\bar{D}^1+i\bar{D}^2)|N(p^i)\rangle$$
$$= \bar{u}_P(p^f)\,((m_P+m_N)\,G_{1,PN}^W(\Delta^2)+\Delta^2 G_{3,PN}^W(\Delta^2))\,\gamma^5 u_N(p^i). \quad (5.5)$$

Keeping only the pion pole in (5.3) we have for small Δ^2

$$(m_P+m_N)\,G_{1,PN}^W(\Delta^2)+\Delta^2 G_{3,PN}^W(\Delta^2)$$
$$\approx -\frac{(f_\pi\cdot\cos\vartheta\cdot m_\pi^2)\,(\sqrt{2}g)}{m_\pi^2-\Delta^2} \qquad (5.6)$$

The $(PN\pi^+)$ coupling constant is conventionally expressed as $(\sqrt{2}g)$. In (5.6) we choose $\Delta^2=0$ and eliminate the Cabibbo

†Measurements of the pion lifetime give $|f_\pi\cdot\cos\vartheta|$.

angle by introducing the axial vector coupling constant $g_A = G^W_{1,PN}(0)/\cos \vartheta$. This gives the Golberger–Treiman formula in its conventional form

$$- (m_P + m_N) g_A \approx f_\pi \cdot \sqrt{2} g \tag{5.7}$$

It is satisfied to within 10 per cent by the independently measured values:

$$g_A \approx 1\cdot18 \pm 0\cdot02; \quad |f_\pi| \approx 0\cdot95 m_\pi; \quad |g| \approx 13\cdot5 \pm 0\cdot1. \tag{5.8}$$

In this and the following chapter we will use pion pole dominance in the form (5.3) to estimate unknown matrix elements of the axial vector divergence $\bar{D}^i$, not only for single-particle states but quite generally. We will use the following prescription[318]: first integrate over internal degrees of freedom in intermediate states and then represent the resulting matrix element by its pion pole. This prescription is unambiguous because we disperse in the total momentum transfer and leave invariant the internal structure of the interacting configuration. For the dominant resonances this prescription can be justified because resonances may effectively be treated like particles; the approximation is more uncertain for the many-particle background, but these states do not provide the dominant contribution in sum rules.

It remains to be noted that the name PCAC (partially conserved axial current) for this pion pole dominance principle is historical (see Chapter 6).

5.2. The Adler–Weisberger Relation

The most impressive test of current algebras is the Adler–Weisberger relation. Its success stimulated the intensive research into current algebras. Here we will derive the relation with the dispersion theory of Section 4.1; therefore we start with the commutator of an axial charge and an axial current:

$$[\bar{Q}^1 + i\bar{Q}^2, \bar{J}^1_\mu(x) - i\bar{J}^2_\mu(x)] = [\bar{Q}^{(+)}, \bar{J}^{(-)}_\mu(x)] = 2 j^3_\mu(x) \tag{5.9}$$

between protons of equal momentum. Equation (4.7) gives

$$2 \langle P | j_\mu^3(0) | P \rangle = 2p_\mu = -\lim_{q \to 0} \int e^{iqx} \vartheta(x_0) \langle P(p) |$$

$$[\bar{D}^{(+)}(x), J_\mu^{(-)}(0)] | P(p) \rangle (dx)^4. \quad (5.10)$$

Taking for the protons the spin average, we have only two kinematic forms

$$2p_\mu = -\lim_{q \to 0} [U^{(1)}(\nu,q^2)p_\mu + U^{(2)}(\nu,q^2)q_\mu]. \quad (5.11)$$

Similarly we decompose

$$u_\mu(\nu,q^2) = \int e^{iqx} \langle P(p) | [\bar{D}^{(+)}(x), \bar{j}_\mu^{(-)}(0)] | P(p) \rangle (dx)^4$$

$$= u^{(1)}(\nu,q^2)p_\mu + u^{(2)}(\nu,q^2)q_\mu. \quad (5.12)$$

The dispersion relation follows the pattern of (4.13) and (4.14)

$$2 = \frac{i}{2\pi} \int \frac{d\nu' u^{(1)}(\nu',0)}{\nu'}. \quad (5.13)$$

By making a further partial integration in $u_\mu(\nu,q^2)$ we can rewrite $u^{(1)}(\nu',0)$:

$$u_\mu(\nu,q^2) = \int e^{iqx} \langle P(p) | [\bar{D}^{(+)}(0), \bar{j}_\mu^{(-)}(-x)] | P(p) \rangle (dx)^4$$

$$iq_\mu u_\mu(\nu,q^2) = i\nu u^{(1)}(\nu,q^2) + iq^2 u^{(2)}(\nu,q^2)$$

$$= \int e^{iqx} \langle P(p) | [\bar{D}^{(+)}(x), \bar{D}^{(-)}(0)] | P(p) \rangle (dx)^4$$

$$\overset{\text{def}}{=} v(\nu,q^2). \quad (5.14)$$

For $q^2 = 0$ we substitute (5.14) into (5.13):

$$2 = \frac{1}{2\pi} \int \frac{d\nu' v(\nu',0)}{(\nu')^2} \quad (5.15)$$

$$v(\nu', 0) = (2\pi)^4 \sum_n \frac{\{\langle P | \bar{D}^{(+)}(0) | n \rangle \langle n | \bar{D}^{(-)}(0) | P \rangle \delta^4(p + q' - p^n)}{(2\pi)^3 2E_n}$$

$$\frac{- \langle P | \bar{D}^{(-)}(0) | n \rangle \langle n | \bar{D}^{(+)}(0) | P \rangle \delta^4(p^n + q' - p)\}}{(2\pi)^3 2E_n} \quad (5.16)$$

63

In (5.15) and (5.16) we separate the neutron contribution, making use of the small proton–neutron mass difference, as explained in Section 4.1. For convenience we take the proton at rest. In the notation of (5.5)–(5.7) the neutron contribution to (5.16) reads:

$$-(2\pi)(m_P+m_N)^2\bar{u}_P\gamma^5 u_N(p_N)\bar{u}_N(p_N)\gamma^5 u_P|g_A|^2\frac{\delta(m_P+q_0'-E_N)}{2E_N}$$

$$=(2\pi)(m_P+m_N)^2\bar{u}_P(\gamma p_N-m_N)u_P|g_A|^2\frac{\delta(m_P+q_0'-E_N)}{2E_N}. \tag{5.17}$$

To give zero momentum transfer the neutron momentum is taken to be

$$p_N=\left(\frac{m_N^2+m_P^2}{2m_P},\frac{m_N^2-m_P^2}{2m_P}e\right)$$

with some unit vector $\vec{e}$.

Between proton spinors at rest only the energy component contributes to (5.17) and thus we have (using invariant normalization: $\bar{u}_P u_P=2m_P$)

$$(2\pi)|g_A|^2(m_N^2-m_P^2)^2\frac{\delta(m_P+q_0'-E_N)}{2E_N}. \tag{5.18}$$

Substituting (5.18) into (5.15), the factor $(m_N^2-m_P^2)^2$ is taken up by

$$\frac{1}{(\nu')^2}=\frac{4}{(m_N^2-m_P^2)^2}$$

and we get with (4.29)

$$2=2|g_A|^2+\int_{\text{continuum}}\frac{d\nu'\;v(\nu',0)}{(\nu')^2}. \tag{5.19}$$

As before, we split the sum over continuum states in (5.16) into an integration over internal degrees of freedom, total mass and spatial momentum. We introduce the invariant functions

64

$$H^{\pm}(W) = \sum_{\text{INT}} \delta(m_n - W) \langle P|\bar{D}^{\pm}(0)|n\rangle_{\Delta^2=0} \langle n|\bar{D}^{\mp}(0)|P\rangle_{\Delta^2=0}$$

and obtain

$$v(v',0) = 2\pi \int\limits_{(m_P+m_\pi)}^{\infty} dW \frac{\delta(m_P + q'_0 - E_W)}{2F_W} \quad (H^+(W) - H^-(W)).$$
$$(5.20)$$

We substitute (5.20) into (5.19), using $(v')^2 = (W^2 - m_P^2)^2/4$ (see (4.25)) and the integral (4.29):

$$1 = |g_A|^2 + \int\limits_{(m_P + m_\pi)}^{\infty} \frac{dW}{(W^2 - m_P^2)^2} (H^+(W) - H^-(W)). \qquad (5.21)$$

Equation (5.21) is exact, in fact it coincides with (2.39), obtained from the charge commutator $[\bar{Q}^{(+)},\bar{Q}^{(-)}] = 2Q^3$ by the infinite-momentum method. Now we make an approximation: because the lepton scattering data for (2.39) are not available we disperse the matrix elements $\langle P|\bar{D}^{\pm}|n\rangle\langle n|\bar{D}^{\mp}|P\rangle$ in $H^{\pm}(W)$ in the momentum transfer q^2 and keep only the pion pole contribution as an approximation.

$$H^{\pm}(W) \approx \sum_{\text{INT}} |f_\pi|^2 \langle P\pi^{\mp}|t^+|n\rangle\langle n|t|P\pi^{\mp}\rangle \delta(m_n - W). \quad (5.22)$$

Up to the factor $|f_\pi|^2$, (5.22) is the transition probability of a $(P\pi)$ system into any state of invariant mass W. This quantity is easily related to the total $(P\pi)$ cross-section at relativistic energy $s = W^2$. We just have to take account of the state normalizations

$$H^{\pm}(W) \approx |f_\pi|^2 \sigma_{\pi^{\mp}P}^{\text{tot}}(W^2) \frac{4W m_P p_\pi^{\text{L}}}{\pi} \qquad (5.23)$$

where p_π^{L} is the (laboratory) momentum of the pion in the proton rest system. So (5.21) takes the form

$$1 = |g_A|^2 + |f_\pi|^2 \int\limits_{(m_P+m_\pi)^2} \frac{ds}{(s - m_P^2)^2} (\sigma_{\pi^-P}^{\text{tot}}(s)$$
$$-\sigma_{\pi^+P}^{\text{tot}}(s)) \cdot \frac{2p_\pi^{\text{L}} \cdot m_P}{\pi} \qquad (5.24)$$

or substituting the Goldberger–Treiman relation (5.7) and replacing $s - m_P^2 = 2m_P E_\pi^{\text{L}} + m_\pi^2 \approx 2m_P E_\pi^{\text{L}}$:

$$1 = |g_A|^2 \left[1 + \frac{2m_P^2}{g^2} \cdot \frac{1}{\pi} \cdot \int\limits_{m_\pi}^{\infty} \frac{p_\pi^L dE_\pi^L}{(E_\pi^L)^2} (\sigma_{\pi^- P}^{\text{tot}}(E_\pi^L) - \sigma_{\pi^+ P}^{\text{tot}}(E_\pi^L)). \right] \quad (5.25)$$

The dominant N^* (1236) resonance produces a negative sign of the integral, so $|g_A| > 1$. With N^* alone we would find $|g_A| \approx 1\cdot4$, but higher isospin ($\frac{1}{2}$) contributions depress the prediction to $|g_A| \approx 1\cdot16$, which compares favourably with the experimental value $1\cdot18 \pm 0\cdot02$.

The derivation of the Adler–Weisberger relation by dispersion theory has been given by Fubini, Furlan and Rossetti.[136] Adler[9] used the infinite momentum limit, Weisberger[385] gave a rather involved derivation for arbitrary proton momenta.

The only approximation is pion pole dominance. In the literature there are different suggestions as to which is the most appropriate place for this approximation. Adler[9,10] connects (5.21) to the scattering of virtual zero mass pions at protons and uses an interesting model to show that the mass shell corrections are small. His corrected result is $|g_A| = 1\cdot24 \pm 0\cdot03$. Weisberger[386] gives arguments for applying the pion pole approximation to the amplitude at $v = 0$, with the neutron Born term separated, and then disperses in v. This suggestion is based on the analyticity properties of $U(v, q^2)$ in both variables. (In Section 6.3 we will use similar considerations when deriving pion scattering lengths.) Schroer and Stichel[350] have derived the Adler–Weisberger relation by dispersion methods based on the Jost–Lehmann–Dyson representation. For some further comments on the Adler–Weisberger relation see references (69) and (214).†

5.3. Generalized Adler–Weisberger Relations

From the commutator (5.9) one can easily derive sum rules for pion total cross sections on any target. To evaluate them is more difficult, because pion scattering has been measured directly only for nucleons and some nuclei. In the case of nuclei it is rather risky to assume pion pole dominance for matrix elements of the axial vector current divergence, because a pion mass is a large energy compared to any nuclear level spacing which may be taken as the energy scale of nuclear dynamics. But experiments should easily provide a test whether Adler–Weisberger sum rules with nuclei[224] are feasible or not.‡

†Recent contributions: (450), (475).   ‡Recent contributions (459), (461), (487).

When considering Adler–Weisberger sum rules for other baryons and mesons as targets one may try to estimate the cross-sections with Breit–Wigner forms for the known resonances and use Regge poles or other models for the high energy contributions. An example is the $(\pi\pi)$ scattering sum rule:

$$2 = |f_\pi|^2 \int\limits_{(2m_\pi)^2}^{\infty} \frac{ds}{(s-m_\pi^2)^2} \left(\sigma^{\text{tot}}_{\pi^+\pi^-}(s) - \sigma^{\text{tot}}_{\pi^+\pi^+}(s)\right) \frac{2p_\pi^{\text{L}} \cdot m_\pi}{\pi}. \quad (5.26)$$

The known $(\pi\pi)$ resonances P and f^0 give about 30–40 per cent of the required value for the integral (5.26). The integrand is heavily weighted towards the threshold, so it is unlikely that high energy contributions will provide the remaining 60 per cent. It may be that the assumption of pion pole dominance introduces uncontrolled errors in this case: a pion mass is the scale of the dynamics in $(\pi\pi)$ scattering and might be too large an energy to extrapolate. Adler[10] has investigated the problem, as far as it is possible to investigate it, and he concludes that such a large error is unlikely. The most natural mechanism to increase the value of the integral would be a strong isospin zero s-wave $(\pi\pi)$ interaction. Such an interaction has been postulated also from the analysis of other processes as the σ-meson.[s 20] The σ-meson may exist as a resonance at around 380 MeV with a large width $\Gamma > 100$ MeV; alternatively it may be only a substitute to parametrize a strong non-resonant $(\pi\pi)$ interaction in the s-wave with isospin zero. Current algebras can only demonstrate the need for such an interaction; one cannot, with a small number of testable sum rules, completely specify its nature. However, if one does assume a σ-resonance, current algebras give satisfactory values for its parameters.[152]

A similar situation is found for the $(K\pi)$ system; the disputed $\kappa(725\ ?)$ meson is too small in width to provide the required 0^+ contribution. Pion scattering sum rules have been derived and estimated by resonance contributions for many targets, in particular for: π;[10,213,215,275] K;[215,253,275] Σ;[96] Ξ;[96,203] P;[130,157-58,252] $K*$;[252] $N*$[362].

Generalized Adler–Weisberger relations can also be constructed for K-meson scattering. Their accuracy is probably more limited, because the assumption of K-meson pole dominance

$$\langle a|\bar{D}^K|b\rangle \approx -\frac{\langle a|t|bK\rangle\langle K|\bar{D}^K|0\rangle}{m_K^2 - \Delta^2} \tag{5.27}$$

proposed in analogy to pion pole dominance (5.3), may introduce larger errors.† In the complex Δ^2-plane the threshold of the cut starts at $(m_K + 2m_\pi)^2 \approx 2m_K^2$, only twice as far away from the origin as the pole at m_K^2. If the distance may be used as an estimate we should expect possible errors up to 50 per cent. A Goldberger–Treiman relation such as (5.7) is not available for a test of the accuracy of (5.27) because the K-meson Yukawa coupling constants to baryons are not known sufficiently well. Thus in applying (5.27) one can only hope for favourable conditions.

In any case the accuracy of (5.27) seems sufficient to demonstrate that the smallness of the measured Cabibbo angle cannot be due to SU_3 violations, if current algebras are correct. Repeating the derivation of the Adler–Weisberger relation for K-meson scattering from the commutator $[\bar{Q}^{K+}, \bar{Q}^{K-}] = \frac{3}{2}Y + I^3$ between protons and neutrons one arrives at

$$2 = |g_A^{P\Lambda}|^2 + |g_A^{P\Sigma^0}|^2 + |f_K|^2 \int\limits_{(m_\Lambda + m_\pi)^2}^{\infty} \frac{ds}{(s-m_P^2)^2}(\sigma_{K^-P}^{tot}(s)$$

$$- \sigma_{K^+P}^{tot}(s))\frac{2p_K^L \cdot m_P}{\pi} \tag{5.28}$$

$$1 = |g_A^{N\Sigma^-}|^2 + |f_K|^2 \int\limits_{(m_\Lambda + m_\pi)^2}^{\infty} \frac{ds}{(s-m_N^2)^2}(\sigma_{K^-N}^{tot}(s)$$

$$- \sigma_{K^+N}^{tot}(s))\frac{2p_K^L \cdot m_N}{\pi} \tag{5.29}$$

The K-meson decay constant f_K and the hyperon axial vector coupling constants are defined with reference to the currents of the algebra, for instance $\langle 0|(\bar{j}_\mu^4(0) - i\bar{j}_\mu^5(0))|K^+\rangle = if_K p_\mu^K$; in the decay of the K-meson one measures $(f_K \cdot \sin\vartheta)$; using $\vartheta \approx 0.212$ we derive $f_K \approx 1.26 f_\pi$. Similarly the hyperon axial vector

†See also reference S23.

coupling constants are derived from the measured quantities by separating a factor $\sin \vartheta$.

Suppose now our separation of the Cabibbo angle as a weak interaction effect outside current algebras had been a misinterpretation of what is in fact an SU_3 violation. Then one would not expect

$$[\bar{Q}^4 + i\bar{Q}^5, \bar{Q}^4 - i\bar{Q}^5] \stackrel{\text{def}}{=} [\bar{Q}^{K+}, \bar{Q}^{K-}] = \tfrac{3}{2}Y + I^3$$

to hold with our definition of $\bar{Q}^K$ (Cabibbo angle separated) but with the measured charges $\bar{Q}^K_* = \bar{Q}^K \sin \vartheta$. If we had

$$[\bar{Q}^{K+}_*, \bar{Q}^{K-}_*] = \tfrac{3}{2}Y + I^3,$$

it would follow that

$$[\bar{Q}^{K+}, \bar{Q}^{K-}] = (\tfrac{3}{2}Y + I^3)/\sin^2 \vartheta \approx 22\,(\tfrac{3}{2}Y + I^3).$$

Such a factor should be detectable even with the limited accuracy of (5.27). In the evaluation of (5.28) and (5.29) an error of about 20 per cent is unavoidable at present, because there are unphysical contributions below threshold in the energy range $(m_\Lambda + m_\pi)^2 \leqslant s \leqslant (m_P + m_K)^2$. These may be approximated by the resonances $Y_0^*(1405)$ and $Y_1^*(1385)$ as effective kaon–nucleon bound states or by complex scattering lengths fitted to low-energy kaon–nucleon scattering and continued below threshold. Although the errors may spoil a quantitative deduction of the hyperon axial vector coupling constants they cannot account for a factor of 22. Thus we conclude that the value of the Cabibbo angle cannot be explained as being due to violations of SU_3 symmetry. For quantitative estimates and discussions of (5.28) and (5.29) see references (23), (47), (241), (302), (347), (386); for other K-meson sum rules see references (71), (213), (251), (253).

In more detailed calculations the errors are unfortunately rather crucial. In principle (5.28) and (5.29) could be used to fit the hyperon decay constants by a (D/F) ratio. In the SU_3 symmetry model the coupling of an octet of axial currents between baryons is fixed by two parameters:

$$\langle B^i | \bar{j}^j | B^k \rangle = (Fif^{ijk} + Dd^{ijk}) \times (\text{kinematic forms}). \quad (5.30)$$

In detail we would have

$$|g_A^{PN}|^2 = (F+D)^2; \quad |g_A^{P\Lambda}|^2 = \frac{3}{2}\left(F+\frac{1}{3}D\right)^2;$$

$$|g_A^{P\Sigma^-}|^2 = 2|g_A^{P\Sigma^\circ}|^2 = (F-D)^2. \quad (5.31)$$

Quantitatively equations (5.28) and (5.29) are rather sensitive to variations of the integrals; an estimate for D/F reflects this sensitivity:[23,386]

$$1\cdot 85 < D/F < 5. \qquad (5.32)$$

An attempt to connect different SU_3 violations[319] by supposedly exact generalized Adler–Weisberger relations is also reduced to a qualitative level by the large errors in present estimates. For $\pi^\pm$ scattering at $K^\pm$ targets or $K^\pm$ scattering at $\pi^\pm$ targets one can derive[253]

$$1 = |f_\pi|^2 \int\limits_{(m_K+m_\pi)^2}^{\infty} \frac{ds}{(s-m_K^2)^2}(\sigma_{K+\pi-}^{\mathrm{tot}}(s) - \sigma_{K+\pi+}^{\mathrm{tot}}(s))\frac{2p_\pi^{\mathrm{L}}.m_K}{\pi} \qquad (5.33)$$

$$1 = |f_K|^2 \int\limits_{(m_K+m_\pi)^2}^{\infty} \frac{ds}{(s-m_\pi^2)^2}(\sigma_{\pi+K-}^{\mathrm{tot}}(s) - \sigma_{\pi+K+}^{\mathrm{tot}}(s))\frac{2p_\pi^{\mathrm{L}}.m_K}{\pi}. \qquad (5.34)$$

There are no resonances known for the $(K^+\pi^+)$ system; so the integrands are positive in the dominating low-energy region. As a consequence of $m_K > m_\pi$ we find from (5.33) and (5.34) that $|f_K| > |f_\pi|$. Since the integrals in (5.28) and (5.29) are positive, any increase in f_K leads to a suppression of the baryon axial vector couplings. This effect also leads to a suppression of the K-meson Yukawa coupling constants if we assume generalized Goldberger–Treiman relations (derived from (5.27)). Such a result is in qualitative agreement with experiment. We may take these arguments as a demonstration of how small departures from SU_3 symmetry must be related to each other to guarantee conservation of exact commutators and we accept as a confirmation of current algebras that the observed pattern emerges. A quantitative test is not yet possible.

Finally we refer to some more general considerations of Adler–Weisberger type relations. Cheng and Kim[95] have systematically investigated their saturation with resonances and found that most sum rules are saturated to about 60–70 per cent by resonant states with masses below 2GeV treated in a single-particle or narrow-width approximation; the rest has to be supplied by continuum states or high-energy contributions. Gilman and Schnitzer[175] have related the high energy parts of different sum rules to each other, using quark model predictions of cross-section differences at high energies. Use has also been made of Johnson–Treiman cross-section relations (from SU_6 and higher symmetries) to obtain information about the continuum part of Adler–Weisberger relations (348, 377).†

5.4. A Photoproduction Sum Rule for the Nucleon Magnetic Moments

In deriving the Adler–Weisberger relation we have used only the matrix elements of the current divergences. This is a general feature of sum rules which can be derived from charge-commutators (or from charge-current commutators if the external states have equal spatial momentum). Between states of equal spatial momentum, matrix elements of current time components can be rewritten:

$$\langle a\,|\,j_0(0)\,|\,b\rangle = \frac{\langle a\,|\,D(0)\,|\,b\rangle}{i(E_a - E_b)}. \tag{5.35}$$

Axial current divergences can be approximated by their meson poles; a similar approximation of vector current divergences is rather hypothetical since no scalar mesons have so far been clearly established. To make further tests on current algebras we have to take charge-current commutators between states of different momenta and derive sum rules into which also the transverse current components enter (these are not determined by the current divergence). The only current whose matrix elements we can hope to know for higher intermediate states is the electromagnetic current. Forming its commutators with pion charges $\bar{Q}^i (i = 1,2,3)$, we will derive sum rules on pion photoproduction. An example of this class will be the subject of this section: a sum rule due to Fubini, Furlan and Rossetti.[138] We consider the commutator

†Recent contributions to this section: (419), (441), (453), (469), (473), (489).

$$[\bar{Q}^i, j_\mu^j] = 0 \quad (i = 3, j = 3, 8) \tag{5.36}$$

between protons and apply the representation (4.7)

$$\lim_{q_{2\mu} \to 0} U_\mu(\nu, t, q_1^2, q_2^2)$$

$$= \lim_{q_{2\mu} \to 0} \int e^{iq_2 x} \langle P(p_2) | [\bar{D}^i(x), j_\mu^j(0)] | P(p_1) \rangle \vartheta(x_0) (dx)^4$$

$$= \lim_{q_{2\mu} \to 0} \int e^{iq_1 x} \langle P(p_2) | [\bar{D}^i(0), j_\mu^j(-x)] | P(p_1) \rangle \vartheta(x_0) (dx)^4. \tag{5.37}$$

Performing a partial integration in the second form of (5.37) we see that $q_1^\mu U_\mu = 0$, using $\partial^\mu j_\mu^j(x) = 0$ and $[\bar{D}^3, \int j_0^j(x)(dx)^3] = 0$ for $j = 3, 8$ with the known quantum numbers of the axial vector current. [In fact this result is rigorously true only after we have taken artificial mass differences to zero; see Section 4.1.] In any case we will not lose essential information if we multiply (5.37) by a constant vector ϵ^μ with $\epsilon^\mu q_{1\mu} = 0$. For $q_2^2 = m_\pi^2$, (5.37) has the form of a pion electroproduction amplitude in LSZ-representation (see Appendix); ϵ^μ is then the polarization vector of the photon.

As U_μ is an axial vector, we have six independent kinematic forms for

$$\epsilon^\mu U_\mu = \sum_{i=1}^{6} \bar{u}_P(p_2) M^{(i)} U^{(i)}(\nu, t, q_1^2, q_2^2) u_P(p_1).$$

They can be taken from papers on pion electroproduction[S 21]

$$M^{(1)} = -\frac{i}{2} \gamma^5 \gamma_\mu \gamma_\nu F^{\mu\nu}$$

$$M^{(2)} = -2i\gamma^5 (p_1 + p_2)_\mu q_{2\nu} F^{\mu\nu}$$

$$M^{(3)} = -\gamma^5 \gamma_\mu q_{2\nu} F^{\mu\nu}$$

$$M^{(4)} = -2\gamma^5 \gamma_\mu (p_1 + p_2)_\nu F^{\mu\nu} + im\gamma^5 \gamma_\mu \gamma_\nu F^{\mu\nu}$$

$$M^{(5)} = -i\gamma^5 q_{1\mu} q_{2\nu} F^{\mu\nu}$$

$$M^{(6)} = -\gamma^5 q_{1\mu} \gamma_\nu F^{\mu\nu}$$

with $F_{\mu\nu} = q_{1\mu}\epsilon_\nu - q_{1\nu}\epsilon_\mu$. The first four are also used in pion photoproduction.[S22] We fix $q_2^2 = 0$, $t = q_1^2$. At the end of the calculation we will consider the limit $t \to 0$ and therefore we will keep only the terms of lowest order in t. ($t \neq 0$ is used only to separate the kinematic forms in (5.38)). For the first invariant matrix element we will consider an unsubtracted dispersion relation:

$$0 = u^{(1)}(0,t,t,0) = \frac{1}{2\pi i} \int \frac{u^{(1)}(\nu',t,t,0)}{\nu'} d\nu'. \qquad (5.39)$$

$u^{(1)}(\nu',t,t,0)$ is the coefficient of $M^{(1)}$ in the decomposition of the discontinuity $u(\nu',t,t,0)$ of $U_\mu \cdot \epsilon^\mu$:

$$u(\nu',t,t,0) =$$

$$(2\pi)^4 \sum_n \left\{ \frac{\langle P(p_2)|\bar{D}^i(0)|n\rangle\langle n|\epsilon^\mu j_\mu^j(0)|P(p_1)\rangle\delta^4(p_2+q_2'-p_n)}{(2\pi)^3 2E_n} \right.$$

$$\left. - \frac{\langle P(p_2)|\epsilon^\mu j_\mu^j(0)|n\rangle\langle n|\bar{D}^i(0)|P(p_1)\rangle\delta^4(p_n+q_2'-p_1)}{(2\pi)^3 2E_n} \right\}. \qquad (5.40)$$

Contributions to (5.40) will come from the proton intermediate state (where we have to use the device of a fictitious mass difference–see Section 4.1) and from a continuum starting at $(m_P + m_\pi)$. We introduce

$$\langle P(p_2)|j_\mu^j(0)|P(p_n)\rangle = \bar{u}_P(p_2)(\gamma_\mu F_1^j(q_1^2)$$

$$+ \sigma_{\mu\nu}(p_n - p_2)_\nu F_2^j(q_1^2))u_P(p_n) \qquad (5.41)$$

with the isovector and isoscalar magnetic moments

$$F_2^3(0) = \frac{1}{2}(\mu_P^A - \mu_N^A), \quad F_2^8(0) = \frac{\sqrt{3}}{2}(\mu_P^A + \mu_N^A) \quad \text{and}$$

$$\langle P(p_2)|\bar{D}^i(0)|P(p_n)\rangle = \frac{-i}{2}g_A(m_P + m_P')\bar{u}_P(p_2)\gamma^5 u_P(p_n). \qquad (5.42)$$

A tedious manipulation of the γ-matrices is necessary to select the coefficient of $M^{(1)}$ in (5.40). In lowest order of t it turns out to be proportional to the anomalous magnetic moment $F_2^j(0)$; this contribution must be balanced by the continuum integral in (5.39). We estimate the integral by assuming pion pole domin-

73

ance in $\langle P(p_2)|\bar{D}^i(0)|n\rangle$ and $\langle n|\bar{D}^i(0)|P(p_1)\rangle$ to introduce transition matrix elements $\langle P\pi^i|t^+|n\rangle$ and $\langle n|t|P\pi^i\rangle$. Equation (5.40) can then be estimated by pion photoproduction data in the limit $q_1^2 = t = 0$ with the help of resonance models[S 30]: the photoproduction data have been analyzed using direct-channel (πN) resonances, with effective coupling constants at vertex (1) to be fitted to the data. The couplings of vertex (2) are known

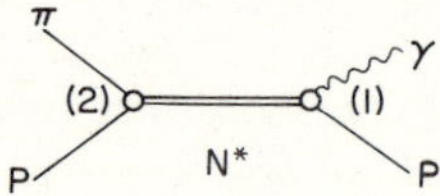

from the resonance decay width. We will also approximate the continuum in (5.40) by the (πN) resonances and use the empirically fitted coupling constants to determine

$$\langle n|j_\mu^j(0)|P(p_1)\rangle \quad \text{and} \quad \langle P(p_2)|j_\mu^j(0)|n\rangle.$$

The dominant N^* (1236) resonance contributes only in the sum rule for the isovector part of the electromagnetic current. Neglecting higher states we would deduce from (5.39)

$$F_2^8(0) = 0 \Rightarrow \mu_P^A = -\mu_N^A \tag{5.43}$$

$$|\mu_P| \approx 2 \text{ nuclear magnetons} \quad \text{(by photoproduction data).} \tag{5.44}$$

This is not very far from the experimental values $\mu_P^A = 1.79$; $\mu_N^A = -1.91$ (nuclear magnetons.)

In resonance models an overall sign of the couplings at the vertices (1) and (2) remains undetermined; this is why we can only make a statement on $|\mu_P|$ and why an inclusion of the next higher resonance $N^*(1515)$ is ambiguous in the relative sign of its contribution. Nevertheless, one can show that the results (5.43) and (5.44) are changed by only 10 per cent. Taking the best combination of relative signs one would arrive at an improved prediction

$$\mu_P = 1\cdot81; \quad \mu_N = -1\cdot99 \quad \text{(nuclear magnetons).} \tag{5.45}$$

The agreement is certainly impressive and justifies the hope that a proper treatment of the full continuum will eliminate the remaining discrepancies. Here we have followed Fubini, Furlan and Rossetti;[138] similar work has been done by Riazuddin and B. W. Lee.[337]

The extension of this calculation for magnetic moments of hyperons has been made by Mathur and Pandit.[254–5] Experimental information is available on K-meson photoproduction in resonance models but in addition a number of unknown couplings have to be taken from the SU_3 symmetry model. As with the nucleons, important contributions come from the $(\frac{3}{2})^+$ decuplet wherever it enters the sum rules. The resulting magnetic moments seem to agree with the rather uncertain experimental values. Their deviations from the SU_3 symmetry predictions are probably not very significant since SU_3 has been used several times in their derivation, where no better approximation was available.†

Despite these good results there is a certain difficulty in all of these calculations. If one considers only the dominant decuplet contributions one finds that the anomalous magnetic moments of the hyperons have to vanish in the SU_3 symmetry limit. To obtain the magnetic moment μ_Λ we might repeat the analysis of this section with the commutator $[\bar{Q}^8, j_\mu^8] = 0$ between Λ particles. The Λ intermediate state will give a contribution proportional to $(g_{A8}^{\Lambda\Lambda} \cdot \mu_\Lambda^A)$. However, the Y_1^* resonance cannot be connected to Λ-states by the isoscalar operators $\bar{Q}^8$ and j_μ^8, so in the decuplet dominance approximation $g_{A8}^{\Lambda\Lambda} \cdot \mu_\Lambda = 0$; there is no d-coupling either for magnetic moments or for axial vector currents. The second alternative is again inconsistent with decuplet dominance, as shown in Chapter 7; so only $\mu_\Lambda = 0$ remains, consequently $\mu_N = 0$ (no d-coupling) and $\mu_P = 0$ (5.43). For the moment we take the least drastic conclusion that the decuplet does not always dominate the continuum part of the integrals in (5.39). Thus in the calculation of μ_Λ one cannot neglect contributions from the Y_0^* (1405) resonance.‡ We will discuss this problem again in Sections 7.3 and 9.5.

Similarly one can use commutators $[\bar{Q}^i, j_\mu^{el}] = -\bar{j}_\mu^i (i = 1,2)$ between nucleons to determine their axial vector form factors in terms of their electromagnetic form factors and pion photoproduction and electroproduction amplitudes.[147,337] The com-

†Recent contribution: (458).   ‡See also (421).

mutators of axial charges and currents $[\bar{Q}^i, \vec{\jmath}^j_\mu] = ie^{ijk}j^k_\mu(i,j,k=1,2,3)$ between nucleons can be used to predict weak (N^*N) axial vector couplings[12,17,146,337] in the approximation of keeping only the nucleon and the $N^*(1236)$ as intermediate states.† These predictions may soon be subject to tests in neutrino experiments.‡ A more refined analysis of the photoproduction sum rules has recently been presented by Adler and Gilman[14] demonstrating the limitations of the narrow-resonance models of previous authors. We also mention a calculation of the induced pseudoscalar form factor in muon capture by similar methods.[235]

The assumption of decuplet dominance and in particular N^* dominance in charge-current commutators should not be taken too far. As we shall see in Section 7.3 the decuplet cannot dominate all dispersion relations, some of them are quite sensitive to the precise mass values[17] and should be treated with particular care. A consideration of charge current commutators with N^* as external state[28] would probably require the inclusion of higher intermediate states.

Assuming the dominance of certain particle and resonance states in sum rules leads to a prediction of their couplings. The power of such predictions as tests of current algebras is rather limited, since a failure of a sum rule might just imply the omission of an important contribution. But if one accepts current algebras as correct, one can use such sum rules as information about the hadron spectrum. In the next section we will give an example of a sum rule which requires the existence of axial vector mesons for its saturation.

5.5. On the Existence of Axial Vector Mesons

If we find an unsatisfactory result in testing current algebra sum rules with a selection of supposedly dominant intermediate states, we have considerable freedom in adding further conjectured states. Such cases are of interest only if the properties of contributing states are sufficiently limited, so that we can use the sum rules to obtain information about the hadron spectrum.

Mesons are so far established with the quantum numbers $J^P = 0^-$, 1^- and 2^+. The discussion about possible 0^+ and 1^+

---

†The weak vector coupling constants in (N^*N) can be taken from N^* photoproduction by CVC.

‡Recent contribution: (496).

mesons is still in progress. We will now give a current algebra sum rule (163,320) which admits only contributions in the channel $J^P = 1^+$. We cannot predict axial vector mesons as single resonances, although this would be the simplest solution; we can only demonstrate the importance of their contributions.

We take the commutator:

$$[\bar{Q}^3, \bar{J}_\mu^+(x)] = j_\mu^+(x) \quad (\text{where } j_\mu^+ = j_\mu^1 + ij_\mu^2) \qquad (5.46)$$

between $\langle \rho^+|$ and the vacuum. Following the general exposition of Section 4.3 we get

$$\langle \rho^+|j_\mu^+(0)|0\rangle \stackrel{\text{def}}{=} f_\rho \epsilon_\mu^\rho = -\lim_{q\to 0} U_\mu(\nu, q^2)$$
$$= -\lim_{q\to 0} \int \vartheta(x_0) e^{iqx} \langle \rho^+|[\bar{D}^3(x), \bar{J}_\mu^+(0)]|0\rangle (dx)^4, \quad (5.47)$$

$$U_\mu(\nu, q^2)$$
$$= U^{(1)}(\nu, q^2)\epsilon_\mu^\rho + U^{(2)}(\nu, q^2)(\epsilon^\rho \cdot q)p_\mu^\rho + U^{(3)}(\nu, q^2)(\epsilon^\rho \cdot q)q_\mu,$$
$$(5.48)$$

$$f_\rho = -U^{(1)}(0,0) = \frac{i}{2\pi} \int \frac{u^{(1)}(\nu', 0)}{\nu'} d\nu', \qquad (5.49)$$

where $u^{(1)}(\nu', 0)$ has to be projected from

$$u_\mu(\nu', 0) = (2\pi)^4 \sum_n \frac{\langle \rho^+|\bar{D}^3(0)|n\rangle \langle n|\bar{J}_\mu^+(0)|0\rangle \delta^4(p^\rho + q' - p_n)}{(2\pi)^3 2E_n}.$$
$$(5.50)$$

Only 0^- and 1^+ mesons can contribute in (5.50). The obvious candidate is the π^+-meson, but we have

$$\langle \rho^+|\bar{D}^3(0)|\pi^+\rangle \sim (\epsilon^\rho p^\pi) = (\epsilon^\rho q') \qquad (5.51)$$

$$\langle \pi^+|\bar{J}_\mu^+|0\rangle \sim p_\mu^\pi = (p_\mu^\rho + q_\mu'). \qquad (5.52)$$

Thus the π-meson intermediate state will contribute only to $u^{(2)}(\nu', 0)$ and $u^{(3)}(\nu', 0)$, but not to $u^{(1)}(\nu', 0)$. Higher pseudoscalar configurations will also not contribute, at least to the extent that they can be treated like single particles. The sum rule (5.49) would therefore lead to the incorrect result $f_\rho \approx 0$, unless there are substantial contributions in the 1^+-channel.

If we had a dominant resonance $|A^+\rangle$, then we would obtain with

$$\langle A^+|\bar{j}_\mu^+|0\rangle = f_A \epsilon_\mu^A \text{ and } \langle \rho^+|\bar{D}^3(0)|A^+\rangle = (\epsilon^\rho \cdot \epsilon^A)G_{\rho^+A^+\pi^\circ} \cdot f_\pi/\sqrt{2}:$$

$$(5.53)$$

$$f_\rho \approx \frac{-i}{\sqrt{2}} \cdot \frac{f_A \cdot G_{A^+\rho^+\pi^\circ} \cdot f_\pi}{M_A^2 - M_\rho^2}.$$

In the same approximation one can show $|f_A| \approx |f_\rho|$, so

$$|f_\pi \cdot G_{A^+\rho^+\pi^\circ}| \approx |\sqrt{2}(M_A^2 - M_\rho^2)|. \tag{5.54}$$

Quantitatively (5.54) would favour a resonance near the ρ-meson mass, but we stress that the equation is based on the possibly wrong assumption that there is a single resonance dominating the 1^+-channel. Current algebra gives us only an integral over 1^+-contributions in (5.49). It does not specify their structure.†

Related sum rules can be constructed which require scalar meson contributions.[163]

†Recent contributions: (436), (509).

CHAPTER 6

Low-Energy Theorems

6.1. General Considerations

In Chapter 4 we related current algebra sum rules to dispersion integrals by considering the function

$$U(\nu,t,q_1^2,q_2^2) = \int \vartheta(x_0) e^{iq_2x} \langle p_2 | [D^i(x), h^j(0)] | p_1 \rangle (dx)^4. \quad (6.1)$$

$U(0,t,t,0) = -c^{ijk}\langle p_2|h^k(0)|p_1\rangle$ is given by current algebras and represented by a dispersion integral. For $\langle p_2|h^k(0)|p_1\rangle$ and for single-particle contributions in the dispersion integral we used weak interaction data whereas the remaining continuum integral was usually approximated by meson pole contributions. Even with this approximation only a few sum rules could actually be estimated.

There exists another class of deductions from current algebras, namely the low-energy theorems. If there are particles of masses m_D and m_h with $\langle 0|D^i(0)|p_D\rangle \neq 0$ and $\langle p_h|h^j(0)|0\rangle \neq 0$, then $U(\nu,t,q_1^2,q_2^2)$ has poles at $q_1^2 = m_h^2$ and $q_2^2 = m_D^2$ with residues $(-i)\langle 0|D^i(0)|p_D(q_2')\rangle\langle p_D(q_2'),p_2|h^j(0)|p_1\rangle$ and $(-i)\langle p_2|D^i(0)|p_1, p_h(q_1')\rangle\langle p_h(q_1')|h^j(0)|0\rangle$. Sometimes the residues are known for physical values of (ν,t) and one may try to approximate the current algebra prediction for $U(0,t,t,0)$ by poles in q_1^2 and q_2^2. This will be illustrated with several examples in the following sections. Three remarks must be made:

(a) In the applications of Chapter 5 we used pole representations mainly in matrix elements $\langle p|\bar{D}^i(0)|n\rangle$ where, at least for dominant contributions, $|n\rangle$ was a single-particle state or a resonance; there we could only write a dispersion integral over the total momentum transfer $(p_p - p_n)^2$ (see (5.3)). This time we have to decide which variables should be kept constant in the analytic continuation of q_1^2 or q_2^2. In most cases it is rather difficult to justify any particular choice.

(b) An approximation to $U(\nu,t,q_1^2,q_2^2)$ by poles in q_1^2 or q_2^2 is reliable only if there is no strong dependence on other variables in the domain of interest. Sometimes we will separate from $U(\nu,t,q_1^2,q_2^2)$ strongly varying Born terms and use meson pole dominance for the rest. Effectively this is what we did in the Adler–Weisberger relation, though we presented the arguments differently because it was the sum rule rather than the low-energy value that was of interest.

(c) If $U(\nu,t,q_1^2,q_2^2)$ contains more than one kinematic form, a separate treatment is necessary for each. Current algebras do not give information about all matrix elements.

An uninitiated reader may ask from where we have our knowledge about the detailed form of the pole residues in (6.1). The Fourier transform of a retarded commutator has essentially the same analyticity properties as a scattering amplitude, which in LSZ representation may be written (see Appendix, equation (A.14)):

$$\langle p_2,q_2|S|p_1,q_1\rangle = -\int e^{iq_2x}\,e^{-iq_1y}(\Box_x+\mu_2^2)\,(\Box_y+\mu_1^2)\{\vartheta(x_0-y_0)$$
$$\langle p_2|[\phi^2(x),\phi^1(y)]|p_1\rangle\}(dx)^4(dy)^4$$

$$=-(2\pi)^4\delta^4(p_1+q_1-p_2-q_2)\int e^{iq_2z}(-q_2^2+\mu_2^2)$$
$$(-q_1^2+\mu_1^2)\{\vartheta(z_0)\langle p_2|[\phi^2(z),\phi^1(0)]|p_1\rangle\}(dz)^4.$$

$$(6.2)$$

$\phi^1(y)$ and $\phi^2(x)$ are the Heisenberg field operators of particles with masses μ_1 and μ_2 and momenta q_1 and q_2. Evidently the integral (6.2) has poles at $q_1^2=\mu_1^2$ and $q_2^2=\mu_2^2$ and the factors $(-q_2^2+\mu_2^2)(-q_1^2+\mu_1^2)$ project the residues; only these have a direct physical meaning. In Lagrangian field theory the integral (6.2) is also defined for arbitrary values of q_1^2 and q_2^2; one introduces the concept of a scattering amplitude for particles off their mass-shell. Perturbation theory suggests that off-shell amplitudes have similar analytic properties to physical scattering amplitudes, at least for a certain neighbourhood.

In Lagrangian field theory Heisenberg operators are introduced as the basic objects and they are specified by equations of motion. If we do not want to take over this formalism we may still use reduction formulas of the type (6.2) for scattering matrix elements provided we find a local operator ψ which can replace the Heisenberg field operator and serve as a so-called interpolating field. This operator must have the quantum numbers of the particle in question, its matrix elements must have a pole at the particle mass with its transition matrix elements in the residue. For a pion, for instance,

$$\langle a|\psi|b\rangle = \frac{\langle a|t|b\pi\rangle \cdot (\text{const})}{m_\pi^2 - \Delta^2} + F(\psi,\Delta^2) \qquad (6.2\text{A})$$

is required with $F(\psi,m_\pi^2)$ finite. Evidently the axial vector divergence $\bar{D}^i$ satisfies this condition through equation (5.3) (which may be regarded as a consequence of generalized unitarity[S 4]). The mere existence of the pion pole in $\bar{D}^i$ and the correct factorization of the pole residue is sufficient; dominance of the pole over other singularities is not yet required at this stage. So we can write reduction formulas for pion scattering amplitudes with the pion field operator ϕ_π^i replaced by a multiple of the axial divergence. We define the substitution by the so-called PCAC equation

$$\phi_\pi^i(x) = \frac{\sqrt{2}}{f_\pi \cdot m_\pi^2}\bar{D}^i(x). \qquad (6.3)$$

The factor is necessary for normalization

$$\langle 0|\phi_\pi^i(0)|\pi^i\rangle = 1; \quad \langle 0|\bar{D}^i(0)|\pi^i\rangle = \frac{1}{\sqrt{2}}f_\pi \cdot m_\pi^2.$$

There is some confusion in the literature about (6.3). Many authors call it an assumption though in our view it is a definition of $\phi_\pi^i(x)$. The assumptions involved in the use of (6.3) will come at a later stage.

Using $\bar{D}^i$ as an interpolating pion field in (6.2) one will get the correct physical scattering amplitudes on the pion mass shell; but it will entirely depend on the general properties of $\bar{D}^i$ what one will get off the pion mass shell. To illustrate the principle assume that there is another pseudoscalar operator E^i (if nothing else is available think of $E^i = (\Box + m_x^2)^N \bar{D}^i$ with $m_x \neq m_\pi$) which also has a pion pole with a correctly factorizing residue

$$\langle a|E^i|b\rangle = \frac{\langle a|t|b\pi\rangle \cdot (\text{const})}{m_\pi^2 - \Delta^2}$$

$$+ (\text{any other singularities at } \Delta^2 \neq m_\pi^2). \qquad (6.4)$$

Instead of (6.3) one can equally well use a suitable multiple of E^i as an interpolating pion field. We will get the correct pion mass-shell scattering amplitudes, but totally different results off the mass shell.

In current algebra applications we can make use only of such interpolating fields whose off-shell scattering amplitudes are dominated by the pion pole:

$$T(q^2) \approx \frac{t(m_\pi^2)}{m_\pi^2 - q^2} \qquad (6.4)$$

where $t(m_\pi^2)$ is the physical transition amplitude. If (6.4) holds we can test a current algebra prediction for $T(q^2)$ by the measured quantity $t(m_\pi^2)$. Using the axial vector divergence as interpolating pion field, the pole dominance principle of Chapter 5 guarantees (6.4). Pion pole dominance is the important assumption now, not (6.3) which is only a definition.

This situation is often presented in a misleading form in the literature; for a clarifying discussion we refer to the papers of Baker[38] and Weinberg.[383] In this discussion we refrained from using involved mathematical methods; there are general theorems by Haag and Nishijima[S 24] demonstrating that the correspondence between particles and fields is not one to one in the LSZ formalism and giving precise conditions for the classes of operators that can be used as interpolating fields.

A final remark should be made about PCAC consistency conditions.[7,8,141−2] How could a definition (6.3) imply physically useful results? Knowing about the existence of an axial vector current whose matrix elements satisfy generalized unitarity, form factor dispersion relations, etc., (6.3) remains a definition. But considering strong interactions only, it is far from obvious that such a regular axial vector current should exist, and it is well conceivable that certain conditions have to be satisfied. This problem may deserve further investigation.

6.2. The *K*-meson Leptonic Decays

As a first example of a low-energy theorem we will derive the relation of Callan and Treiman[90] on the *K*-meson leptonic decays. We consider the commutator

$$[\bar{Q}^3, j_\mu^4(x) - ij_\mu^5(x)] = [\bar{Q}^3, j_\mu^{K^-}(x)] = -\tfrac{1}{2}j_\mu^{K^-}(x) \qquad (6.5)$$

between the vacuum and $|K^+\rangle$;

$$U_\mu(\nu = p^K \cdot q, q^2) = \int \vartheta(x_0)e^{iqx}\langle 0|[\bar{D}^3(x), j_\mu^{K^-}(0)]|K^+\rangle$$

$$= U^{(1)}(\nu, q^2)p_\mu^K + U^{(2)}(\nu, q^2)q_\mu. \qquad (6.6)$$

The equal-time commutator gives $\langle 0|j_\mu^{K^-}(0)|K^+\rangle \overset{\text{def}}{=} if_K p_\mu^K$, so we have $U^{(1)}(0,0) = \tfrac{1}{2}if_K$ and no statement on $U^{(2)}(0,0)$. $U_\mu(\nu, q^2)$ has a pole at $q^2 = m_\pi^2$ with the residue

$$(-i)\langle 0|\bar{D}^3(0)|\pi^0\rangle \cdot \langle \pi^0|j_\mu^{K^-}(0)|K^+\rangle = \frac{-i}{\sqrt{2}} \cdot f_\pi \cdot m_\pi^2 \langle \pi^0|j_\mu^{K^-}(0)|K^+\rangle$$

in terms of the $\pi^\pm$-decay constant f_π. ($f_{\pi^0} = f_\pi/\sqrt{2}$ by isospin invariance). $\langle \pi^0|j_\mu^{K^-}|K^+\rangle$ can be measured in the decay $K^+ \rightarrow \pi^0 + \mu^+ + \nu$:

$$\langle \pi^0|j_\mu^{K^-}(0)|K^+\rangle = \frac{-1}{\sqrt{2}}\left[F_{K\pi}^+(\Delta^2)(p_\mu^K + p_\mu^\pi) + F_{K\pi}^-(\Delta^2)(p_\mu^K - p_\mu^\pi)\right]$$

$$= \frac{-1}{\sqrt{2}}[(F_{K\pi}^+(\Delta^2) + F_{K\pi}^-(\Delta^2))p_\mu^K + (F_{K\pi}^+(\Delta^2)$$
$$- F_{K\pi}^-(\Delta^2))p_\mu^\pi].$$
$$(6.7)$$

In the pole residue $(F_{K\pi}^+(\Delta^2) + F_{K\pi}^-(\Delta^2))$ contributes to $U^{(1)}(\nu, q^2)$, $(F_{K\pi}^+(\Delta^2) - F_{K\pi}^-(\Delta^2))$ to $U^{(2)}(\nu, q^2)$. Δ^2 is connected to ν by $\Delta^2 = m_K^2 - 2\nu + q^2$. The momentum transfer in the pole residue will depend on which variable we keep fixed; fortunately this dependence is not critical since the rate of change in the form factors $F_{K\pi}^+(\Delta^2)$ and $F_{K\pi}^-(\Delta^2)$ is given by the squared masses of vector mesons and possible scalar mesons, which are much larger than m_π^2. So we can deduce

$$U^{(1)}(0,0) = \tfrac{1}{2} i f_K \approx \tfrac{i}{2} f_\pi [F_{K\pi}^+(m_K^2) + F_{K\pi}^-(m_K^2)]. \tag{6.8}$$

In SU_3 symmetry $F_{K\pi}^-(\Delta^2) \equiv 0$ and $F_{K\pi}^+(0) = 1$. Experimentally only $(f_K . \sin \vartheta)$, $(f_\pi . \cos \vartheta)$ and $(F_{K\pi} . \sin \vartheta)$ are measured. [In fact by measuring $(F_{K\pi}^+(0) \sin \vartheta)$ and using the Ademollo–Gatto theorem[S 15] to set $F_{K\pi}^+(0) = 1$ one obtains the best possible estimate of the Cabibbo angle.] Equation (6.8) is free from this uncertainty, because we can multiply it by $(\sin \vartheta)$; $(f_\pi . \cos \vartheta)$ is not sensitive to errors in ϑ. The uncertainties are introduced through the unknown dependence of $F_{K\pi}^+(\Delta^2)$ and $F_{K\pi}^-(\Delta^2)$ on the momentum transfer, and in particular through the errors in the not easily accessible quantity $\xi = F_{K\pi}^-/F_{K\pi}^+ = 0.3 \pm 0.3$[S 31]. Using a K^* pole model for $F_{K\pi}^+$:

$$F_{K\pi}^+(m_K^2) \approx F_{K\pi}^+(0) \frac{(m_K{}^*)^2}{(m_K{}^*)^2 - m_K^2}$$

and neglecting the dependence of $F_{K\pi}^-(\Delta^2)$ on Δ^2 we find:

$$f_K - \sin \vartheta / f_\pi \approx F_{K\pi}^+(0) \left[\frac{(m_K{}^*)^2}{(m_K{}^*)^2 - m_K^2} + \xi \right] . \sin \vartheta$$

$$0.27 \approx [1.45 + (0.3 \pm 0.3)]0.21 = 0.37 \pm 0.06. \tag{6.9}$$

A more accurate determination of the K-meson decay parameters would be desirable. So far we can only say that (6.9) is in qualitative agreement with experiment. In deducing (6.8) we used only pion pole dominance. We might also have started from the commutator

$$[\bar{Q}^{K^0}, j_\mu^{K^-}(x)] = \bar{J}_\mu^{\pi^-}(x)$$

between vacuum and $|\pi^+\rangle$ and using the more disputed hypothesis of K-meson pole dominance in matrix elements of the divergence $\bar{D}^K$ (equation (5.27)) we would get

$$f_\pi / f_K \approx [F_{K\pi}^+(m_\pi^2) - F_{K\pi}^-(m_\pi^2)]. \tag{6.11}$$

84

On general grounds we would not expect (6.11) to be very accurate though with the present data it seems to be well satisfied: $0.78 = 0.7 \pm 0.3$. A qualitative connexion between two SU_3 violations seems to be required by (6.9) and (6.11): $f_K/f_\pi > 1$ and $F_{\bar{K}\pi}^-(0) > 0$. The smallness of ξ also shows us that we should expect $f_K \approx f_\pi$. Returning to the problems of Section 1.3, equations (1.23) and (1.24), we see that Cabibbo[s 12] has (perhaps intuitively?) chosen the correct scaling to compare the meson decay constants with the SU_3 symmetry model. Whatever his arguments were, the SU_3 symmetry model does not give us information about the best scaling; it is one of the advantages of current algebras that here we do not have any scaling ambiguities.

The connections of $K(l3)$ and $K(l2)$ decays† as described above have been studied by several authors: references (90), (150), (207), (251), (256), (288), (336), (344), (354). The current algebra results (6.8) and (6.11) have been combined with form factor dispersion relations for $F_{K\pi}^+(\Delta^2)$ and $F_{K\pi}^-(\Delta^2)$, which again were approximated by meson poles. For the details we refer to the original papers given above; we only warn the uninitiated reader that in some papers approximations are used whose accuracy is not yet clearly established.

Applying similar techniques to the comparison of $\pi(l3)$ and $\pi(l2)$ decays leads back to the predictions of the CVC theory, discussed in Section 1.2. This connection has been investigated under different aspects in the references (56), (90), (204), (322).

Very interesting is the consideration of the decay $K(l4)$: $K \rightarrow 2\pi + \text{leptons}$, mediated mainly through the weak axial vector current. The two π-mesons are in an s-wave state; any reasonable approximation should maintain their symmetry. One possible treatment would be a successive reduction. We could start from

$$U_\mu^{(\alpha\beta)}(\nu,t,q_1^2,q_2^2) = \int \vartheta(x_0) e^{iq_2 x} \langle \pi^{(\alpha)}(p_2) | [\bar{D}^{(\beta)}(x), J_\mu^{K^-}(0)] | K^+(p_1) \rangle$$

$$(dx)^4 \qquad (6.12)$$

and approximate the current algebra prediction on $U_\mu(0,t,t,0)$ by the pion pole in $q_2^2 = m_\pi^2$ with the desired matrix element $\langle \pi^{(\alpha)}(p_2),$

---

† $K(l3)$: leptonic K-decay into three particles, etc.

$\pi^{(\beta)}(q'_2)|\bar{J}_\mu^{K^-}|K^+(p_1)\rangle$ in its residue. By adding the corresponding result from $U_\mu^{(\beta\alpha)}$ and by taking $t = m_K^2$ we impose the symmetry between the pions.

Another way would maintain the symmetry by starting from integrals over time-ordered products:

$$\int (dx)^4 (dy)^4 e^{iq_2 x} e^{ip_2 y} \langle 0| T(\bar{D}^{(\alpha)}(y), \bar{D}^{(\beta)}(x), \bar{J}_\mu^{K^-}(0)) E|K^+(p_1)\rangle (dx)^4 \tag{6.13}$$

The analytic continuations in (6.12) and (6.13) for which pion pole dominance is to be postulated are not identical; this will reflect itself in the appearance of so-called σ-terms (which we will discuss in the next section for a simpler example). Another problem in $K(l4)$ calculations is the choice of suitable kinematic forms. The discussion about $K(l4)$ decays is still in progress; we prefer to refer to the literature here,[53, 68, 90, 100, 206, 367, 382], rather than give a possibly one-sided view of this rather complex problem.

Before developing new concepts we should try to understand what kind of problem can be investigated by the methods of this section. Effectively we related a matrix element $\langle a, \pi^i|$ $j_\mu(0)|b\rangle$ in the pole residue of U_μ to the commutator $\langle a|[\bar{Q}^i, j_\mu(0)]|b\rangle$ as the low-energy limit of U_μ. Such connections have been known as Ward–Takahashi identities,[316-317] but current algebras are necessary to make a testable statement by proposing a definite form for the commutator. So far we have definite ideas about these commutators only if j_μ is a vector or axial-vector current. Consequently we can make deductions only on leptonic decays into one or several pions and on pion production by photons and neutrinos; in principle one may also attempt to describe K-meson production. The subject of pion production has been treated in the following papers: references (35), (231), (277), (295), (316), (317), (374). It is worth noting that the Kroll–Ruderman theorem[S 25] can be reproduced from current algebras: references (231), (295), (374). A consideration of gauge invariance in the low-energy theorems on pion production by Nauenberg[277] may be of more general significance.†

6.3. Scattering Lengths of Mesons

In this section and in the following one we describe a second class of low-energy theorems. We start again with a remarkable

†Recent contributions to this section: (393), (418), (423), (474), (499).

example and will afterwards investigate the range of generalizations. We consider the commutator of two axial isocharges and currents

$$[\bar{Q}^i, \bar{J}^j_\mu(x)] = ie^{ijk} j^k_\mu(x) \qquad (i,j,k = 1,2,3) \tag{6.14}$$

between nucleons at rest, taken in the spin average for simplicity.

$$ie^{ijk}\langle p| j^k_\mu(0)|p\rangle = -\lim_{q\to 0} \int \vartheta(x_0) e^{iqx}\langle p|[\bar{D}^i(x), \bar{J}^j_\mu(0)]|p\rangle (dx)^4$$

$$= -\lim_{q\to 0} \int \vartheta(x_0) e^{iqx}\langle p|[\bar{D}^i(0), \bar{J}^j_\mu(-x)]|p\rangle (dx)^4. \tag{6.15}$$

We multiply (6.15) by iq^μ and perform a partial integration:

$$0 = \lim_{q\to 0} \int \vartheta(x_0) e^{iqx}\langle p|[\bar{D}^i(0), \bar{D}^j(-x)]|p\rangle (dx)^4$$

$$-\lim_{q\to 0} \int \delta(x_0) e^{iqx}\langle p|[\bar{D}^i(0), \bar{J}^j_0(-x)]|p\rangle (dx)^4. \tag{6.16}$$

We introduce

$$V(\nu, q^2) = \int \vartheta(x_0) e^{iqx}\langle p|[\bar{D}^i(x), \bar{D}^j(0)]|p\rangle (dx)^4 \tag{6.17}$$

and rewrite equation (6.16)

$$0 = V^{ij}(0,0) - \langle p|[\bar{D}^i, \bar{Q}^j]|p\rangle \overset{\text{def}}{=} V^{ij}(0,0) - \sigma^{ij}. \tag{6.18}$$

The second term is the commutator of an axial charge with an axial divergence. Its value is unknown except in special models from which it received the name σ-term. In (6.18) the protons have equal spatial momentum, therefore σ^{ij} is already determined by a charge commutator:

$$[\int \bar{D}^i(x)(dx)^3, \bar{Q}^j] = \left[\frac{d}{dt}\bar{Q}^i, \bar{Q}^j\right] = \frac{d}{dt}[\bar{Q}^i, \bar{Q}^j] - \left[\bar{Q}^i, \frac{d}{dt}\bar{Q}^j\right]$$

$$= \frac{d}{dt}[\bar{Q}^i, \bar{Q}^j] + \left[\frac{d}{dt}\bar{Q}^j, \bar{Q}^i\right] \tag{6.19}$$

In current algebra theory the commutator $[\bar{Q}^i,\bar{Q}^j] = ie^{ijk}Q^k$ is a conserved isovector charge, so σ^{ij} has to be symmetric in its isospin indices.

$V(\nu,q^2)$ has a double pole in q^2 at m_π^2 with the pion-nucleon scattering amplitude in its residue, taken at energy ν. So we want to find a way to continue from $\nu = 0$, $q^2 = 0$ to physical values $\nu \geqslant m_P \cdot m_\pi$, $q^2 = m_\pi^2$ without uncontrolled variations in $V(\nu,q^2)$. We have to examine the contribution of the nucleon Born pole at $(m')^2 = (p \pm q)^2 = m_P^2 \pm 2\nu + q^2$; as explained in Section 4.1 we keep artificially $m' \neq m_P$. The nucleon Born pole does not contribute in configurations where the intermediate nucleon is at rest relative to the external nucleon. Such a system is an eigenstate of positive parity and therefore cannot be coupled to a pseudoscalar operator $\bar{D}(x)$.

Along the surface $q^2 = 0$ the contribution of the nucleon Born pole is of the order $(m_P - m') \to 0$. This can be shown in detail by considering the discontinuity $v(\nu,0)$ across the real ν-axis with $q^2 = 0$. This has already been done in Section 5.2 when deriving the Adler–Weisberger relation. In equation (5.18) we found a contribution of the order $(m_P - m')^2$ to the discontinuity $v(\nu,0)$. [The contribution to the amplitude $V(\nu,0)$ at $\nu = 0$ is only of the order $(m_P - m')$ because the pole is located at

$$\nu' = \pm \frac{((m')^2 - m_P^2)}{2m_P} \approx \pm (m' - m_P).]$$

Therefore in extrapolating from $\nu = 0$ to threshold $\nu = m_P \cdot m_\pi$ we need not consider the nucleon Born pole. At $\nu = m_P \cdot m_\pi$ we will use pion pole dominance in q^2.

A constant extrapolation $V(m_P \cdot m_\pi,0) \approx V(0,0) = \sigma^{ij}$ is unsatisfactory, because it would imply the equality of $(P\pi^+)$ and $(P\pi^-)$ scattering lengths. So we attempt a linear extrapolation:

$$V(m_P \cdot m_\pi) \approx V(0,0) + (m_P \cdot m_\pi)\frac{\partial V(0,0)}{\partial \nu}. \qquad (6.20)$$

It is characteristic for this class of low-energy theorems that the derivative $\partial V(0,0)/\partial \nu$ can be computed from current algebras, which we have hardly used so far. We go back to equation (4.3), introducing:

$$T_{\mu\nu}(\nu,q^2) = \int \vartheta(x_0)e^{iqx}\langle p|[\,\bar{j}_\mu^j(x),\bar{j}_\nu^i(0)\,]|p\rangle\,(dx)^4, \qquad (6.21)$$

$$U_\nu(\nu,q^2) = \int \vartheta(x_0) e^{iqx} \langle p|[\bar{D}^i(x),\bar{j}^j_\nu(0)]|p\rangle (dx)^4. \quad (6.22)$$

For $q \neq 0$ we obtain:

$$iq^\mu T_{\mu\nu}(\nu,q^2) + U_\nu(\nu,q^2) + \int \delta(x_0) e^{iqx} \langle p|[\bar{j}^i_0(x),\bar{j}^j_\nu(0)]|p\rangle (dx)^4$$
$$= 0. \quad (6.23)$$

We multiply (6.23) by iq^ν and use the partial integration (6.16)

$$-q^\mu q^\nu T_{\mu\nu}(\nu,q^2) + V(\nu,q^2) - \int e^{iqx} \delta(x_0) \langle p|[\bar{D}^i(x),\bar{j}^j_0(0)]|p\rangle (dx)^4$$

$$+ iq^\nu \int e^{iqx} \delta(x_0) \langle p|[\bar{j}^i_0(x),\bar{j}^j_\nu(0)]|p\rangle (dx)^4 = 0. \quad (6.24)$$

Equation (6.24) is covariant if the commutators are sufficiently regular at the origin (as will be assumed; for more singular cases see Chapter 8). In a system where the space component of q vanishes we can use the commutators (6.14) and (6.18) to write:†

$$-q^\mu q^\nu T_{\mu\nu}(\nu,q^2) + V(\nu,q^2) - \sigma^{ij} + iq^\nu i e^{ijk} \langle p|j^k_\nu(0)|p\rangle = 0. \quad (6.25)$$

Differentiating (6.25) with respect to ν, the first term will not contribute at $\nu = 0$, $q^2 = 0$ because it is of second order in q (remember that we avoid all kinds of degeneracy), σ^{ij} is independent of ν (if it is at all properly defined), $\langle p|j^k_\nu|p\rangle$ has the form $c^k P_\nu$ with a constant c^k. So we get:

$$\frac{\partial}{\partial\nu} V(\nu,0) = e^{ijk} c^k. \quad (6.26)$$

Equation (6.26) is the Adler–Weisberger relation in the form of a low-energy theorem; by assuming unsubtracted dispersion relations in ν for $(\partial/\partial\nu)V(\nu,0)$ we could rederive it by standard methods.

Now we take a concrete example: $P\pi^-$ scattering; the commutator in question is $[\bar{Q}^1 + i\bar{Q}^2, \bar{j}^1_\mu(x) - i\bar{j}^2_\mu(x)] = 2j^3_\mu(x)$ for (6.14); with protons we have $c^3 = 2$, so (6.20) gives

$$V(m_P . m_\pi,0) \cong \sigma - 2i(m_P . m_\pi) \quad (6.27)$$

†Equation (6.25) is a covariant one, so it is valid in every system.

and applying pion pole dominance:

$$V(m_P \cdot m_\pi, 0) \approx -f_\pi^2 \cdot \langle P\pi^- | P\pi^- \rangle. \tag{6.28}$$

The threshold value of the scattering amplitude is usually expressed by the scattering length†

$$(a_{P\pi-}) = -i\frac{\langle P\pi^- | P\pi^- \rangle}{8\pi \cdot (m_P + m_\pi)} = \frac{1}{8\pi f_\pi^2}\left[\frac{i\sigma + 2m_P \cdot m_\pi}{(m_P + m_\pi)}\right] \tag{6.29}$$

$$(a_{P\pi+}) = \qquad\qquad = \frac{1}{8\pi \cdot f_\pi^2}\left[\frac{i\sigma - 2m_P \cdot m_\pi}{(m_P + m_\pi)}\right] \tag{6.30}$$

$$(a_{P\pi-}) - (a_{P\pi+}) = \frac{1}{2\pi \cdot f_\pi^2} \cdot \frac{m_P \cdot m_\pi}{(m_P + m_\pi)} \approx 0\cdot23 \text{ fermi;}$$

$$\text{expt } 0\cdot252 \pm 0\cdot007 \text{ fermi.} \tag{6.31}$$

Several authors have derived these results by different continuations from the low-energy value to threshold. In some derivations the σ-term appears, in others not (for instance reference (39)). If we set it equal to zero we would deduce $(a_{P\pi}^-) = -(a_{P\pi}^+) \approx 0\cdot115$ fermi which is satisfied with the same accuracy as (6.31): $(a_{P\pi}^+) = -0\cdot123 \pm 0\cdot007$ fermi, $(a_{P\pi}^-) = 0\cdot129 \pm 0\cdot007$ fermi. The smallness of the σ-term is also suggested by models; we would rather regard it as a practical PCAC consistency condition: if one way of continuation produces the term and a different one does not, then pion pole dominance can be universally valid only if the σ-term is small or zero.‡

In our derivation we have worked in the spin average with equal proton momenta. By considering more complicated configurations, information can be extracted not only about the s-wave scattering lengths, (6.29) and (6.30), but also about certain combinations of p-wave scattering lengths. The possibility of predictions finds a limit in the term $q_\mu q_\nu T_{\mu\nu}(\nu, q^2)$ which physically would correspond to the scattering of axial vector mesons at nucleons; apart from its Born term it is entirely unknown. Thus no quantity can be deduced in this method which

---

†Remember that $\langle P\pi^- | t | P\pi^- \rangle$ is an S-matrix element, hence the factor $(-i)$.
‡See also (466).

is of second order in q. We refer to the literature for further details on pion–nucleon scattering lengths: references (39), (40), (41), (142), (143), (314), (315), (316), (317), (373), (383).

It is very easy to generalize the results to pion s-wave scattering lengths at arbitrary targets. As long as the approximations are applicable, the scattering lengths are determined only by the mass and by the isospin of the target. The linear extrapolation (6.20) is meaningful only if the scattering amplitude is very smooth below threshold. This is well illustrated by the K-meson nucleon system. The prediction for (a_{PK^+}), obtained by setting the σ-term equal to zero comes out quite well despite the large extrapolation:

$$(a_{PK^+}) \approx \frac{1}{4\pi}\,\frac{m_P}{m_P+m_K}\cdot\left(\frac{2m_K}{f^2_K}\right) \approx -0{\cdot}4\,\text{fermi, expt} \approx -0{\cdot}29\,\text{fermi}.$$

There is however no hope of finding the scattering length of the (PK^-) system; in this case one finds below threshold a $|\Lambda\pi)$ and a $(\Sigma\pi)$ cut, containing the Y^*_0 and Y^*_1 resonances.

The pion–nucleon scattering lengths could be determined because the pion mass is not too large compared to the scale of the dynamics. In pion–pion scattering the linear extrapolation fails because it cannot offset the asymmetry in the treatment of the pions. Weinberg[383] has therefore used an expansion in all scalar variables for the off-mass-shell pion–pion scattering; this method preserves crossing symmetry, Bose statistics and isospin invariance and permits an extrapolation to the physical threshold. With a linear expansion Weinberg finds small scattering lengths. By making additional algebraic assumptions Khuri[229] has confirmed this result in a quadratic approximation. Both Weinberg and Khuri admit that their results are reliable only if the pion scattering lengths are really small. They have not disproved the alternative that they might be large due to a low lying s-wave $(\pi\pi)$ resonance $\sigma(?)$; in such a case the approximation scheme would be inadequate. As current algebras require a substantial low-energy s-wave $(\pi\pi)$ interaction (Section 5.3), we would regard the problem as still open.†

6.4. Vector Meson Decays

Before drawing general conclusions about the method of

†Recent contributions to this section: (404), (408), (434), (440), (451), (462).

Section 6.3 we will consider a related low-energy theorem given by Kawarabayashi and Suzuke.[216] We consider the same commutators as in Section 6.3, only different states, vacuum and ρ-meson;

$$T_{\mu\nu} = \int \vartheta(x_0) e^{iq_2 x} \langle 0|[\bar{J}^i_\mu(x), \bar{J}^j_\nu(0)]|\rho(p_1)\rangle$$

$$= \int \vartheta(x_0) e^{-iq_1 x} \langle 0|[\bar{J}^i_\mu(0), \bar{J}^j_\nu(-x)]|\rho(p_1)\rangle$$

$$U_\nu = \int \vartheta\Big((x_0) e^{iq_2 x} \langle 0|[\bar{D}^i(x), \bar{J}^j_\nu(0)]|\rho(p_1)\rangle,$$

$$V = \int \vartheta(x_0) e^{iq_2 x} \langle 0|[\bar{D}^i(x), \bar{D}^j(0)]|\rho(p_1)\rangle.$$

The momenta attached to the axial operators are q_1 and q_2, satisfying $q_1 + q_2 = p_1$. As scalar variables we choose q_1^2 and q_2^2. This time there is no Born pole in the problem, there is also no σ-term because $[\bar{Q}^i, \bar{D}^j(0)]$ is rotationally invariant and cannot connect the vacuum to a ρ-meson. So we can proceed immediately to equation (6.25) with the only modification that we want to treat the two axial operators symmetrically; therefore we average over the two possible orders of partial integration. We obtain:

$$-q_2^\mu q_1^\nu T_{\mu\nu}(q_1^2, q_2^2) - V(q_1^2, q_2^2)$$
$$+\tfrac{1}{2} i (q_1 - q_2)^\nu i e^{ijk} \langle 0|j^k_\nu(0)|\rho(p_1)\rangle = 0. \qquad (6.33)$$

The quantity $V(q_1^2, q_2^2)$ is a pseudoscalar. It must be linear in the polarization vector $\epsilon^{(\rho)}_\mu$ of the ρ-meson. Because of the condition $\epsilon^{(\rho)}_\nu p^\nu_{(\rho)} = \epsilon^{(\rho)}_\nu (q_1 + q_2)^\nu = 0$ there remains only one possible kinematic form: $V(q_1^2, q_2^2) = \epsilon^{(\rho)}_\nu (q_1 - q_2)^\nu G(q_1^2, q_2^2)$. Similarly $\langle 0|j^k_\nu|\rho\rangle = f^\rho \cdot \epsilon^{(\rho)}_\nu$ has a unique form. In the limit $q_1 \to 0$, $q_2 \to 0$ one may compare the two first order terms in (6.33) and deduce

$$G(0,0) = -\tfrac{1}{2} e^{ijk} f^\rho. \qquad (6.34)$$

For $G(q_1^2, q_2^2)$ we may use pion pole dominance and express it by the $(\rho\pi\pi)$ coupling constant in the residue. Taking the example $\rho^\circ \to \pi^+\pi^-$ we derive in detail for (6.34)

$$(f_\pi)^2 g_{\rho\pi\pi} = f^\rho. \qquad (6.35)$$

Equation (6.35) is not directly testable, but f^ρ can also be determined by the assumption that form factors of the conserved

92

isovector current are dominated by an effective ρ-pole. Between pions we have

$$\langle \pi^+(p_1) | j_\mu^3 | \pi^+(p_2) \rangle = F(\Delta^2)(p_1 + p_2)_\mu \quad \text{with} \quad F(0) = 1.$$

In a dispersion relation for $F(\Delta^2)$ the ρ-pole model gives

$$F(\Delta^2) \approx \frac{f^\rho g_{\rho\pi\pi}}{m_\rho^2 - \Delta^2}, \quad \text{in particular} \quad 1 = \frac{f^\rho g_{\rho\pi\pi}}{m_\rho^2}. \tag{6.37}$$

Recently f^ρ has been determined experimentally by the electromagnetic decay $\rho \to \mu^+ + \mu^-$ in agreement with (6.37) though with appreciable errors. With (6.35) and (6.37) we can eliminate f^ρ and keep

$$f_\pi^2 (g_{\rho\pi\pi})^2 = m_\rho^2. \tag{6.38}$$

This leads to a decay width for $\rho \to 2\pi$:

$$\Gamma = \frac{|p_\pi|^3}{6\pi} \cdot \frac{(g_{\rho\pi\pi})^2}{m_\rho^2} \approx 140 \text{ MeV}; \text{ experimentally} \sim 125 \text{ MeV}.$$

Similar calculations can of course be done for K^* decay and for the $\phi \to K\bar{K}$ decay if K-meson pole dominance is used. For the ϕ decay the $\omega - \phi$ mixing angle is necessary as an input. The main idea in this class of low-energy theorems is a relation between matrix elements of transition processes $\langle A\pi^i\pi^j | B \rangle$ (in the residue of the double pion pole) and the matrix element of an isovector current $e^{ijk} \langle A | j_\mu^k | B \rangle$ which arises in first order in the pion momenta. The zeroth order is given by the ambiguous σ-terms, the first order becomes relevant either in a linear extrapolation (Section 6.3) or through favourable kinematic forms (this section). It is important to realize the antisymmetry of the result in the pion isospin indices: current algebra describes only configurations where the pions are in a p-wave, if both are emitted or absorbed, or in a p-wave of the crossed channel if one is emitted and one absorbed (Section 6.3). This has been observed by Sakurai[345] who finds that the results for the nucleon–pion scattering lengths are reproduced if one makes the assumption that ρ- exchange dominates low-energy nucleon-pion scattering and uses (6.38) for the ρ-coupling:

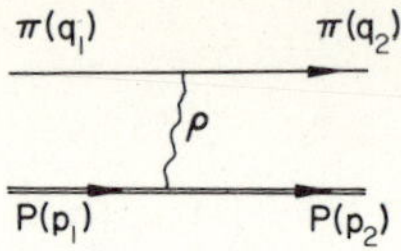

The significance of this connection should be clear by now. The low-energy theorems of this section have the same scope as ρ exchange, provided there are no difficulties with kinematic forms which make their application questionable.

Such cases are found in the decays $(\omega \to 3\pi)/(\omega \to \pi^\circ + \gamma)$ and $(\eta \to 2\pi + \gamma)/(\eta \to 2\gamma)$. In using an equation analogous to (6.25) we find that the term V, for instance,

$$V = \int \vartheta(x_0) e^{iq_2 x} \langle \pi^0(p_2) | [\bar{D}^i(x), \bar{D}^j(0)] | \omega(p_1) \rangle (dx)^4$$

$$= e^{\mu\nu\kappa\lambda} \epsilon_\mu^\omega p_{1\nu} q_{1\kappa} q_{2\lambda} G(\nu, t, q_1^2, q_2^2) \tag{6.39}$$

is of second order in the pion momenta, like the term $q_{1\mu} q_{2\nu} T_{\mu\nu}$. There is now no convincing way to separate the two terms. A similar situation occurs in the η-decays. This has been clearly brought out by Kellett.[223] Neglecting without clear justification the term $q_{1\mu} q_{2\nu} T_{\mu\nu}$, some authors have nevertheless obtained reasonable results.† It now seems clear why they succeeded: this class of low energy theorems is connected to ρ-exchange which is dynamically important in ω-decays and presumably also in η-decays:

Below we will list the contributions in the literature to low-energy theorems of this class. We warn the reader that the understanding of these theorems has gradually developed; not all papers are free from misleading statements: references (2), (3), (100), (119), (173), (205), (216), (223), (228), (256), (305), (309), (335), (345–6), (379), (404), (408), (434), (440), (451), (462), (492).

†See also (492).

By related methods Weinberg has developed a theory of multiple pion production.[381]

6.5. Non-leptonic Weak Decays

Leptonic and semileptonic weak interactions have been described successfully by the coupling of weak currents in the Hamiltonian (1.2). To complete the picture, it would be most natural to have a symmetrized† product of hadron weak currents describing the non-leptonic weak decays.

$$\mathscr{H}_{NL} = \frac{G}{\sqrt{2}} \cdot \frac{1}{2} [J_\mu^{W,h}(J^{\mu W,h})^+ + (J_\mu^{W,h})^+ J^{\mu W,h}]. \qquad (6.40)$$

This form cannot be tested directly because one cannot separate the two currents as in the semileptonic decays

$$\mathscr{H}_{SL} = \frac{G}{\sqrt{2}} [J_\mu^{W,h}(J^{\mu W,l})^+ + (J_\mu^{W,h})^+ J^{\mu W,l}] \qquad (6.41)$$

The indirect evidence is at first sight even against (6.40). Constructed with the Cabibbo current

$$J_\mu^{W,h} = \cos \vartheta \, [\, (j_\mu^1 + i j_\mu^2) + (\bar{j}_\mu^1 + i\bar{j}_\mu^2)\,] + \sin \vartheta \, [\, (j_\mu^4 + i j_\mu^5) + (\bar{j}_\mu^4 + i\bar{j}_\mu^5)\,],$$

$$(6.42)$$

$\mathscr{H}_{NL}$ gives selection rules $|\Delta Y| \leqq 1$ and $|\Delta I_3| \leqq \frac{1}{2}$, $\Delta I = \pm\frac{1}{2}$ or $\pm\frac{3}{2}$ for $\Delta Y = \pm 1$; but it does not explain the empirical $\Delta I = \frac{1}{2}$ rule which states that in the decay of strange particles $\mathscr{H}_{NL}$ acts like an operator with isospin $\frac{1}{2}$. This rule is satisfied in good approximation; one finds, for instance: $\Gamma_{K^+ \to \pi^+ \pi^0} \ll \Gamma_{K^0 \to \pi^+ \pi^-}$ or $\Gamma_{\Lambda \to N\pi^0} \approx \frac{1}{2}\Gamma_{\Lambda \to P\pi}$. In $\mathscr{H}_{NL}$ (6.40) both isospin $\frac{1}{2}$ and isospin $\frac{3}{2}$ components are present; it is not evident *a priori* how the latter ones come to be suppressed. Therefore (6.40) is not yet established and alternative forms, explicitly containing the $\Delta I = \frac{1}{2}$ rule, are still being discussed (see Section 7.4). Current algebra is the first theory to reconcile the Hamiltonian (6.40) with the $\Delta I = \frac{1}{2}$ rule by reproducing its results as low-energy theorems, first derived by Sugawara[360] and Suzuki.[366]

†An antisymmetrized product would have $CP = -1$.

The construction of $\mathcal{H}_{NL}$ from vector and axial vector currents implies the following interesting relation:

$$[Q^i,(j+\bar{\jmath})^j_\mu] = if^{ijk}(j+\bar{\jmath})^k_\mu = [\bar{Q}^i,(\bar{\jmath}+j)^j_\mu]$$

$$[Q^i,\mathcal{H}_{NL}] = [\bar{Q}^i,\mathcal{H}_{NL}]$$

The commutator of $\mathcal{H}_{NL}$ with an axial charge is the same as that of $\mathcal{H}_{NL}$ with the corresponding isospin or SU_3 generator.

(a) Hyperon Decays

To consider the decay of hyperons $B_1 \to B_2 + \pi$ we study the commutator

$$\langle B_2(p_2)|[\bar{Q}^i,\mathcal{H}_{NL}(0)]|B_1(p_1)\rangle$$
$$= -\lim_{q \to 0} \int \vartheta(x_0)e^{iqx}\langle B_2(p_2)|[\bar{D}^i(x),\mathcal{H}_{NL}(0)]|B_1(p_1)\rangle(dx)^4. \quad (6.43)$$

To simplify the treatment we will work in the limit of almost equal baryon masses and small pion mass; this limit is certainly not far from reality. Under the simplifying approximation $p_1 \approx p_2 \approx p$ we can regard the integral in (6.43) as a function of $\nu = p \cdot q$ and q^2 only. For the pseudoscalar part $\mathcal{H}^P_{NL}$ of $\mathcal{H}_{NL}$ the integral is a scalar and will be denoted by $U^{(S)}$; for the scalar part $\mathcal{H}^S_{NL}$ of $\mathcal{H}_{NL}$ the integral is a pseudoscalar: $U^{(P)}$.

$$\langle B_2(p)|[\bar{Q}^i,\mathcal{H}^P_{NL}(0)]|B_1(p)\rangle = -\bar{u}_2(p)U^{(S)}(\nu = 0,q^2 = 0)u_1(p),$$
$$(6.44)$$
$$\langle B_2(p)|[\bar{Q}^i,\mathcal{H}^S_{NL}(0)]|B_1(p)\rangle = -\bar{u}_2(p)U^{(P)}(\nu = 0,q^2 = 0)u_1(p),$$
$$(6.45)$$

At $q^2 = m^2_\pi$ the integral has a pole with the non-leptonic decay amplitude in the residue $(-i)1/\sqrt{2}f_\pi m^2_\pi \langle B_2,\pi^i|\mathcal{H}_{NL}(0)|B_1\rangle$; this can be related to the physical decay amplitude if small variations in $(p_1 - p_2)^2$ and ν are neglected. The current algebra limit $U(0,0)$ is approximated by the physical amplitude in the residue of the pion pole at $q^2 = m^2_\pi$ with the baryon Born poles separated:

For the pseudoscalar part $\mathcal{H}^P_{NL}$, producing the parity violating *s*-wave decays, the baryon Born pole contribution is of order $O(\Delta m)$ in the baryon mass splittings. The situation is analogous to the role of the nucleon Born term in the determination of pion–nucleon scattering lengths. In the limit $q \to 0$ we have two pseudoscalar vertices of order $O(\Delta m)$ and a propagator of order $O(1/\Delta m)$. This is not the case for the parity violating *p*-wave decays produced by the scalar part $\mathcal{H}^S_{NL}$. Though $\langle B_1 | \mathcal{H}^P_{NL} | B_2 \rangle = O(\Delta m)$, $\langle B_1 | \mathcal{H}^S_{NL} | B_2 \rangle = O(1)$ because a scalar operator can have a non-vanishing matrix element between states which are asymptotically at rest with respect to each other. Therefore for *p*-wave decays the Born term will give a finite contribution strongly dependent on q and will not allow a smooth extrapolation of the current algebra limit to a physical pion momentum.

So we first compute from current algebras the matrix element $\langle B_2 | [\bar{Q}^i, \mathcal{H}^P_{NL}] | B_1 \rangle$ and relate it to the parity violating decay amplitudes.

$$\langle B_2 | [\bar{Q}^i, \mathcal{H}^P_{NL}(0)] | B_1 \rangle \approx i \frac{1}{\sqrt{2}} f_\pi \cdot m_\pi^2 \langle B_2 \pi^i | \mathcal{H}^P_{NL}(0) | B_1 \rangle. \qquad (6.46)$$

We cannot evaluate (6.46) absolutely, but we can use the properties of $[\bar{Q}^i, \mathcal{H}^P_{NL}(0)] = [Q^i, \mathcal{H}^S_{NL}(0)]$ under SU_3. $\mathcal{H}_{NL}$ is a symmetric product of two octet currents, so it may have components in a singlet, an octet and a twenty-seven-dimensional multiplet; the same is true for its scalar and pseudoscalar part and their isospin partners in (6.46). The singlet no longer occurs there and the octet and twenty-seven-plet components are just transformed in isospin. Thus we can use the Wigner–Eckart theorem to express the matrix elements $\langle B_2 | [Q^i, \mathcal{H}^S_{NL}(0)] | B_1 \rangle$ by three undetermined parameters, two for $\langle 8|8|8 \rangle$, one for $\langle 8|27|8 \rangle$. This gives us four relations between seven observed decay amplitudes (*s*-wave parts):

$$\sqrt{2}\, T(\Xi^0 \to \Lambda \pi^0) + T(\Xi^- \to \Lambda \pi^-) = 0, \qquad (6.47)$$

$$\sqrt{2}\, T(\Lambda \to N \pi^0) + T(\Lambda \to P \pi^-) = 0, \qquad (6.48)$$

$$T(\Sigma^- \to N \pi^-) - \sqrt{2}\, T(\Sigma^+ \to P \pi^0) - T(\Sigma^+ \to N \pi^+) = 0, \qquad (6.49)$$

$$2\, T(\Xi^- \to \Lambda \pi^-) + T(\Lambda \to P \pi^-) = \sqrt{\tfrac{3}{2}}\, T(\Sigma^+ \to N \pi^+) + \sqrt{3}\, T(\Sigma^+ \to P \pi^0). \qquad (6.50)$$

Assuming that the relatively slow decay $\Sigma^+ \to N\pi^+$ goes mainly through p-waves, the relations (6.47)–(6.50) are compatible with the experimental data. Their predictions differ from those of the $\Delta I = \frac{1}{2}$ rule only by unobservable phases. With (6.47)–(6.50) current algebras have shown the emergence of effective $\Delta I = \frac{1}{2}$ transitions for the hyperon s-wave decays in the limit of SU_3 symmetry and small pion mass.

The p-wave decays receive important contributions from the baryon Born terms; due to their strong momentum dependence we should take here the masses of broken SU_3. The vertices are given by $\langle B_2|\mathscr{H}^S_{NL}|B_n\rangle$ and $\langle B_n|\mathscr{H}^S_{NL}|B_1\rangle$, which can be computed in (6.46) from the s-wave decays, and by the pion-baryon coupling constants, which are not well known and must be taken from SU_3. The p-wave decays cannot be explained by the Born terms only: though one may fit the data to Born graphs with free vertices $\langle B_2|\mathscr{H}^S_{NL}|B_n\rangle$ and $\langle B_n|\mathscr{H}^S_{NL}|B_1\rangle$, the solution obtained for these vertices is not compatible with (6.46). Evidently there have to be further contributions. For details we refer to the literature: references (33), (34), (36), (42), (68), (80), (126), (190), (201), (273), (360), (366) (388).† We regret that in particular the treatments of p-wave decays are somewhat uncorrelated and that current algebra deductions are not always clearly separated from additional assumptions. With similar methods Ω^- decays$^{(188,312)}$ and the decay $\Sigma^+ \to P + \gamma^{(370)}$ have been investigated.

(b) *K-Meson Decays*

As with the hyperon decays there is again an effective $\Delta I = \frac{1}{2}$ rule to be explained; it is observed in $K \to 3\pi$ and $K \to 2\pi$ decays. We consider first $K \to \pi^i + \pi^j$ and use a successive reduction:

$$\langle \pi^i|[\bar{Q}^j, \mathscr{H}_{NL}(0)]|K\rangle = -\lim_{q_2 \to 0} \int \vartheta(x_0) e^{iq_2 x}\langle \pi^i(p_2)|[\bar{D}^j(x),$$
$$\mathscr{H}_{NL}(0)]|K(p_1)\rangle (dx)^4. \qquad (6.51)$$

The commutator on the left-hand side of (6.51) just produces an isospin partner of $\mathscr{H}_{NI}$:

$$\mathscr{H}'_{NL}(0) \overset{\text{def}}{=} [Q^j, \mathscr{H}_{NL}(0)] = [\bar{Q}^j, \mathscr{H}_{NL}(0)];$$

†Recent contributions: (438), (471), (472), (495), (514).

the integral on the right-hand side is approximated by its pion pole at $q_2^2 = m_\pi^2$ with the residue $(-i)(f_\pi/\sqrt{2})\langle\pi^i\pi^j|\mathcal{H}_{NL}|K\rangle$, which we compare to the physical amplitude of $K \to 2\pi$. We do not know $\langle\pi^i|\mathcal{H}'_{NL}(0)|K\rangle$ in (6.51), therefore we reduce again:

$$\langle 0|[\bar{Q}^i, \mathcal{H}'_{NL}(0)]|K\rangle = -\lim_{p_2\to 0} \int \vartheta(x_0)e^{ip_2x}\langle 0|[\bar{D}^i(x),$$
$$\mathcal{H}'_{NL}(0)]|K(p_1)\rangle(dx)^4. \qquad (6.52)$$

As in (6.51) the left-hand side produces again an isospin partner of $\mathcal{H}_{NL}$: $\mathcal{H}''_{NL}(0) = [Q^i, \mathcal{H}'_{NL}(0)] = [\bar{Q}^i, \mathcal{H}'_{NL}(0)]$; the integral will be approximated by its pion pole at $p_2^2 = m_\pi^2$ with the residue $(-i)(f_\pi/\sqrt{2})\langle\pi^i|\mathcal{H}'_{NL}(0)|K\rangle$, just what we need for the left-hand side of (6.51).

If the pion pole approximation to the limit of vanishing pion momenta is reliable (we shall investigate this below), then we have established a proportionality:

$$\langle\pi^i\pi^j|\mathcal{H}_{NL}(0)|K\rangle \sim \langle\pi^i|\mathcal{H}'_{NL}(0)|K\rangle \sim \langle 0|\mathcal{H}''_{NL}(0)|K\rangle. \quad (6.53)$$

In (6.53) we split the Hamiltonian $\mathcal{H}_{NL}(0)$ into an isospin $(\frac{1}{2})$ and $(\frac{3}{2})$ part: $\mathcal{H}_{NL}(0) = \mathcal{H}^{(1)}(0) + \mathcal{H}^{(3)}(0)$. In the commutation with the isospin generators Q^i and Q^j the total isospin of the terms $\mathcal{H}^{(1)}(0)$ is unchanged, only the third component may change. Only $\mathcal{H}^{(1)}(0)''$ can connect a K-meson to the vacuum, therefore

$$\langle\pi^i\pi^j|\mathcal{H}^{(3)}(0)|K\rangle \sim \langle 0|[\mathcal{H}^{(3)}(0)]''|K\rangle = 0. \qquad (6.54)$$

The isospin $(\frac{3}{2})$ part does not contribute to the current algebra limit. For the isospin $(\frac{1}{2})$ part the remarkable observation is made that the current algebra limit depends on the ordering of the axial charges $\bar{Q}^i$ and $\bar{Q}^j$ in the reduction process. We have Jacobi's identity:

$$[\bar{Q}^i[\bar{Q}^j, \mathcal{H}_{NL}(0)]] - [\bar{Q}^j[\bar{Q}^i, \mathcal{H}_{NL}(0)]] = [[\bar{Q}^i, \bar{Q}^j], \mathcal{H}_{NL}(0)]]$$
$$= ie^{ijk}[Q^k, \mathcal{H}_{NL}(0)]. \quad (6.55)$$

To extrapolate to physical amplitudes we have to conserve Bose statistics. Since the final pions are in a spatially symmetric s-wave state they must also be symmetric in their isospin variables. This eliminates the ambiguity of ordering: we have to symmetrize with respect to (i,j).

Before making a final statement we have to examine to what extent the current algebra limit of vanishing meson momenta is reliably approximated by the pion poles, which contain the physical transition matrix element in their residue. We look into possible singularities of the integral (6.51) in other channels using the similarity of its analytic properties to a scattering amplitude:

A spurion of momentum q_1 has been associated to $\mathcal{H}_{NL}$. We

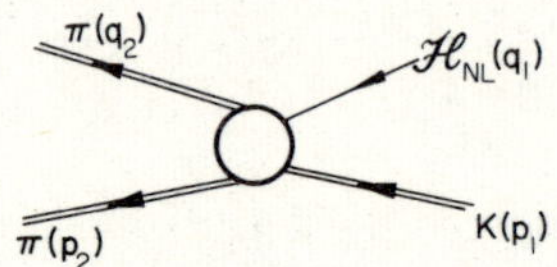

introduce the usual variables $p_1^2 = m_K^2, p_2^2, q_2^2, q_1^2$, $s = (p_1+q_1)^2 = (p_2+q_2)^2$, $t = (p_2-p_1)^2 = (q_2-q_1)^2$, $u = (p_1-q_2)^2 = (p_2-q_1)^2$. The following extrapolations are necessary from the current algebra limit to the physical decay point:

$p_2^2: 0 \rightarrow m_\pi^2$ (pion pole dominance).

$q_2^2: 0 \rightarrow m_\pi^2$ (pion pole dominance).

$t: m_K^2 \rightarrow m_\pi^2$ (uncritical, because there are only slow variations produced by the K^* resonance).

$u: m_K^2 \rightarrow m_\pi^2$ (uncritical, as with t).

$s: 0 \rightarrow m_K^2$ (This extrapolation cannot be justified if there is a σ-resonance, whose mass is presumably less than m_K^2. The isospin of $\sigma(?)$ is zero, so its presence would invalidate the deductions on $\mathcal{H}^{(1)}(0)$, but not on $\mathcal{H}^{(3)}(0)$ which connects the K-meson only to an $I = 2$ configuration of the final pions.)

$q_1^2: m_K^2 \rightarrow 0$ (This channel contains a singularity from the graph:

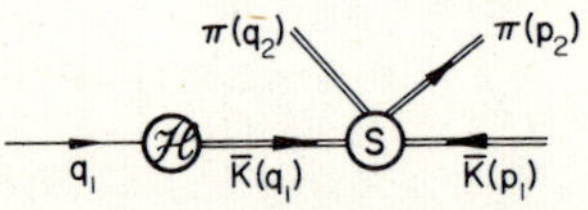

The spurion is connected to a $\bar{K}$-meson by $\langle \bar{K}| \mathcal{H}_{NL}(0)|0\rangle$; the propagator of the $\bar{K}$-meson is singular at $q_1^2 = m_K^2$. Again this graph contributes only with $\mathcal{H}^{(1)}$, not with $\mathcal{H}^{(3)}$.

We summarize: Although we should not attempt to compare the low energy value for the matrix element of $\mathcal{H}^{(1)}$ with the physical decay amplitude, for $\mathcal{H}^{(3)}$ we can safely approximate it by the pion poles at $q_2^2 = m_\pi^2$ and $p_2^2 = m_\pi^2$. We find in (6.54) that the low-energy value vanishes, so we can expect the same for the physical decay: only $\mathcal{H}^{(1)}$ is effective in $K \to 2\pi$ decays; it produces the observed $\Delta I = \frac{1}{2}$ rule.

In the decays $K \to 3\pi$ we have again to explain an empirical $\Delta I = \frac{1}{2}$ rule. It is possible to relate the process $K \to 3\pi$ to $K \to 2\pi$ by current algebras. We consider the integral

$$\langle \pi^i \pi^j | [\bar{Q}^k, \mathcal{H}_{NL}(0)] | K \rangle$$
$$= -\lim_{q \to 0} \int \vartheta(x_0) e^{iqx} \langle \pi^i \pi^j | [\bar{D}^k(x), \mathcal{H}_{NL}(0)] | K \rangle (dx)^4$$
$$\approx \lim_{q \to 0} \frac{i(f_\pi/\sqrt{2}) \cdot m_\pi^2}{m_\pi^2 - q^2} \langle \pi^i \pi^j ; \pi^k | \mathcal{H}_{NL}(0) | K \rangle \qquad (6.56)$$

assuming again pion pole dominance. In (6.56) we certainly introduce errors in neglecting final-state interactions, such as σ-configurations of π^k with π^i and π^j; this error cannot be avoided at present. In a three-body decay it is not necessary to consider spurion channels. We introduce as variables the pion masses $(q^\alpha)^2$ and the energies $((1/m_K)p^K \cdot q^\alpha) = E^\alpha$ in the K-meson rest system: $E^1 + E^2 + E^3 = m_K$. The current algebra limit is given by $(q^1)^2 = (q^2)^2 = m_\pi^2; (q^3)^2 = 0; E^1 = E^2 = \frac{1}{2}m_K$. In order that (6.56) should be reliable we must preserve the symmetry in the isospin indices of the pions. Because of the small momentum release the pions are most likely in an s-wave state with respect to each other; this is a completely symmetrical configuration. Correspondingly we should symmetrize (6.56) with respect to (i,j,k) before comparing it with experiments.

We see that the $K \to 3\pi$ decays are expressed by the isospin $(\frac{1}{2})$ part $\mathcal{H}^{(1)}$ of $\mathcal{H}_{NL}$, because only this one is effective in $K \to 2\pi$ decays. If we neglect any dependence of $\langle 3\pi | \mathcal{H}^{(1)} | K \rangle$ on the pion energies, i.e. if we assume a uniform Dalitz plot, then we express all matrix elements of $\langle 3\pi | \mathcal{H}^{(1)} | K \rangle$ by one free scaling parameter which can be determined from (6.56). Suzuki[368] has predicted the relation:

$$\Gamma(K^+ \to \pi^+\pi^+\pi^-)/\Gamma(K_1^0 \to \pi^+\pi^-) \approx 4\cdot8 \times 10^{-4} \ [\text{expt } 6\cdot3 \pm 0\cdot5]$$
$$(6.57)$$

between the experimentally best known decay modes. By unsymmetric reductions it has also been possible to deduce successfully the slope parameters for the energy distributions of the three pions. The extrapolation procedures from the current algebra limit to physical momenta are not unique among different authors. We do not want to commit ourselves to any particular version here and leave it to the reader to decide which one he may find most convincing. Literature on *K*-meson decays: references (20), (53), (68), (73), (78), (90), (120–1), (189), (264), (279), (368), (389).†

There have been a number of speculations about intermediate vector bosons. Suppose the weak Hamiltonian (1.2) is not completely correct: instead of being coupled to each other (1.2), the weak currents are coupled to a vector boson field W_μ:

$$H = \{f \cdot (J_\mu^W \cdot W^\mu) + f((J_\mu^W)^+ W^{\mu+})\}. \qquad (6.58)$$

If this vector boson field has only a charged form but not a neutral one, this would describe the fact that the weak currents have to be charged. A weak reaction which we used to describe by the product of two weak currents would in fact proceed by *W*-boson exchange; β-decay, for instance, would follow the diagram:

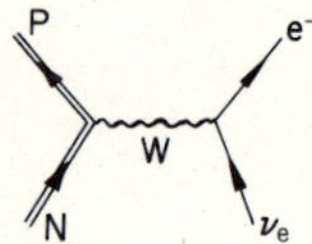

Instead of the familiar form

$$\frac{G}{\sqrt{2}} (J_\mu^{W,h}) (J^{\mu W,e})^+, \qquad (6.59)$$

from (1.2) we would have

$$f^2 (J_\mu^{W,h}) \left(\frac{+g^{\mu\nu} - \Delta^\mu \Delta^\nu / m_W^2}{m_W^2 - \Delta^2} \right) (J_\nu^{W,e})^+ \qquad (6.60)$$

describing the decay. ($\Delta^\mu = (p^p - p^N)^\mu$.) If m_W is very large, then (6.60) will tend to the limiting form (6.59) with $G/\sqrt{2} = +f^2/_W^2$.

†Recent contributions: (390), (420), (452), (482), (495).

102

The picture would not be quite as simple with non-leptonic decays; the W-boson exchange can occur somewhere inside a complicated diagram. In the integrals (6.43), (6.51) and (6.52) $\mathcal{H}_{NL}(0)$ would be replaced by

$$f^2 \int \Delta_{\mu\nu}(z) \left[(J^W_\mu(z)) \cdot (J^{\mu W}(-z))^+ + (J^W_\mu(z))^+ (J^{\mu W}(-z)) \right] (dz)^4$$

where $\Delta_{\mu\nu}(z)$ is a Green function of the W-boson, and the ϑ-function would be replaced by a more general time-ordering operator to preserve covariance. To what extent would this affect the current algebra deductions?

Bell[53] has shown that the low-energy limits are not affected. The extrapolations to physical pion momenta encounter additional singularities due to the non-local nature of the Hamiltonian. As long as $m_W \gg m_\pi$ these additional singularities will not contribute appreciably. But they may contribute and make the extrapolation unreliable when we are dealing with second-order electromagnetic reactions with massless photon exchange. This is well illustrated by a result of Sutherland[363], who finds that the electromagnetic decay amplitude for $\eta \to 3\pi$ will vanish if the energy of one pion is taken to vanish. This result is very interesting and its implications are disputed. We refer to the literature: (75), (363), (376), (392), (420), (426), (439), (452), (490), (500). The authors disagree in their proposals.

Finally we wish to mention a number of papers, (64), (72), (97), (98), (99), (128), (188), (202), where the attempt is made to calculate the matrix elements $\langle B_2 | \mathcal{H}'_{NL}(0) | B_1 \rangle$ and $\langle \pi | \mathcal{H}'_{NL}(0) | K \rangle$ by substituting intermediate states into the (current) $\times$ (current) product of $\mathcal{H}_{NL}(0)$. There is no momentum conservation for intermediate states in the product of two local operators, therefore matrix elements $\langle A | j_\mu | B \rangle$ have to be integrated over all momentum transfers. Only for the electromagnetic current between nucleons are the matrix elements known for large momentum transfer. Keeping the baryon octet and sometimes also the decuplet or alternatively a selection of mesons as intermediate states, the authors have to make very strong assumptions on unknown form factors. The results are of the correct order of magnitude; more should not be expected.

CHAPTER 7

Extensions of the SU$_3$ × SU$_3$ Current Algebra

7.1. The Chiral SU$_6$ × SU$_6$ Algebra

We have found that an $SU_3 \times SU_3$ algebra can be proposed for the electromagnetic and weak currents which leads to many successful predictions. We may ask whether there exist larger but still finite algebras of relevance to physics.

We can, for example, integrate the commutator

$$[Q^i, j_\mu^j(x)] = if^{ijk} j_\mu^k(x) \qquad (7.1)$$

over x for $\mu \neq 0$ and form a space-component charge

$$Q_\mu^i = \int j_\mu^i(x)(dx)^3; \qquad (7.2)$$

similarly with axial currents. But we do not know the commutator of two space-component charges. Before postulating the form of these commutators we will examine their properties in the quark model of Section 2.3. There the currents are bilinear forms in the quark fields: $q^+(x)\Gamma(\lambda_i/2)q(x)$. λ_i specifies the internal quantum numbers, Γ is one of the sixteen hermitean Dirac matrices.

Using a representation $\gamma_0 = \left(\begin{smallmatrix} 1 & \\ & -1 \end{smallmatrix}\right)$; $\gamma_i = \left(\begin{smallmatrix} & -\sigma_i \\ \sigma_i & \end{smallmatrix}\right)$; $\gamma_5 = \left(\begin{smallmatrix} & 1 \\ 1 & \end{smallmatrix}\right)$ we find for the

vector charges   (time component) Q_0^i: $\Gamma = 1$,

                     (space component) Q_k^i: $\Gamma = \left(\begin{smallmatrix} & -\sigma_k \\ -\sigma_k & \end{smallmatrix}\right)$,

axial charges   (time component) $\overline{Q}_0^i$: $\Gamma = \left(\begin{smallmatrix} & -1 \\ -1 & \end{smallmatrix}\right)$,

                    (space component) $\overline{Q}_k^i$: $\Gamma = \left(\begin{smallmatrix} \sigma_k & \\ & \sigma_k \end{smallmatrix}\right)$. $\qquad (7.3)$

Apart from these bilinear forms the quark model also contains tensor currents (six components), scalar currents and pseudoscalar currents. The commutators in the quark model are:

$$[Q^i(\Gamma),Q^j(\Gamma')] = \tfrac{1}{2}if^{ijk}Q^k(\{\Gamma,\Gamma'\}_+) + \tfrac{1}{2}d^{ijk}Q^k([\Gamma,\Gamma']). \qquad (7.4)$$

The algebra of the 144 bilinear forms closes. It contains all hermitean operators in the twelve-dimensional quark-triplet space, so it generates the group U_{12} of all unitary transformations. For the moment the U_{12}-algebra is too large for us because we do not know which physical quantities may correspond to the tensor, scalar and pseudoscalar currents. All we can identify so far are the space and time components of vector and axial vector currents and we have to look for subalgebras of U_{12} which contain only these.

An important subalgebra S is generated by Q_0^i and $\bar{Q}_k^i$ ($i=0\ldots 8$, $k=1,2,3$). As can be seen from (7.3) these generators do not connect upper and lower components of the quark Dirac spinors. In the two invariant six-dimensional subspaces they have the form $\tfrac{1}{2}\lambda_i\sigma_j$, familiar from the SU_6 model (Section 1.4); so S is an SU_6 algebra. The full algebra of (7.3) is $SU_6 \times SU_6$ with generators ($Q_0^i \pm \bar{Q}_0^i$, $Q_k^i \pm \bar{Q}_k^i$).

We now abstract the commutators (7.4) from the quark model and investigate the suggestion that they might be valid also with physical weak currents; this hypothesis was developed mainly by Dashen, Gell-Mann and B. W. Lee:[112],[113],[129],[240],[278]. The early papers on this subject are recommended for their intrinsic interest; in their arguments the concept of approximate symmetry for these algebras (compare Chapter 3 for SU_3) plays a larger role than we would assign to this concept today. This is quite natural because the extended current algebras (in different forms) were developed as an interpretation of the SU_6 symmetry model.

To become more familiar with the $SU_6 \times SU_6$ algebra, we want to study its representations. First we take a model space of single-particle states and later we will specify to what extent these model representations are an approximation to physical reality. In the model we first confine ourselves to the rest frame; this is consistent because the generators, being space integrals, leave the momentum invariant.

Excluding parity doubling, the matrix elements of Q_k^i and $\bar{Q}_0^i$ vanish between states at rest, so only the subalgebra S needs to be represented. The representation spaces must have the dimensionality of an SU_6 group representation and must allow the construction of axial vectors (for $\bar{Q}^i_k$); this would not be possible with only spin zero particles. There is, however, no need to have $\sqrt{\tfrac{3}{2}}\bar{Q}^0_k$ (containing the identity matrix in internal quantum numbers)† equal to the spin operators S_k as in the non-relativistic SU_6 symmetry model. A so-called L-excitation is allowed[167] with $\sqrt{\tfrac{3}{2}}\bar{Q}^0_k = S_k + L_k$. Instructive examples for this point have been worked out.[341]

Sometimes the set of states in question specifies the representation uniquely by its SU_3 and spin multiplets; this is the case for an octet of spin $(\tfrac{1}{2})^+$ combined with a decuplet of spin $(\tfrac{3}{2})^+$, or for an octet of spin 0^- combined with a nonet of spin 1^-. In these examples the matrix elements of the generators have the SU_6 values, familiar from the rest symmetry model: references (127), (240), (291), (341).

The SU_6 predictions $|g_A| = \tfrac{5}{3}$, $(D/F)_{Ax} = \tfrac{3}{2}$ obtained from the model-representation of fifty-six baryon states are physical to the extent that the real baryons and mesons do indeed provide a representation of S. This is the case in the approximation that in commutators taken between the fifty-six baryon states only the same set gives appreciable contributions as intermediate states. (In fact the results for the baryon octet follow already from the weaker assumption that a commutator taken between octet baryons is saturated by the octet and decuplet states, without specifying anything about commutators taken between octet–decuplet or decuplet–decuplet states.) Whether this approximation is good or bad must be left to the dynamics of strong and weak interactions, since the generators are physically already completely defined. Any additional assumption about saturation of commutators by certain states is a dynamical statement and it should be justified dynamically, otherwise there is a danger that it will conflict with previous specifications about current algebras. We will see later that the baryon model is not without a dynamical basis, but it is certainly not accurate, since $|g_A| = 1{\cdot}18$, not $1{\cdot}67$.

Although results of the SU_6 symmetry model can be obtained from current algebras, the postulates are completely different. We do not assume a rest symmetry which cannot be extended to

†Remember $\lambda_0 = \sqrt{\tfrac{2}{3}}$

106

the relativistic domain without making great sacrifices. Instead we postulate an algebra between observables and consider representations under certain approximations: these approximations can be improved, as we shall see in Section 7.2. Even in a quark model with any assignment of masses, the spatial charges are always time-dependent operators. Okubo has shown that a symmetry limit, where the spatial charges become time-independent, does not exist.[296]

As Ryan has shown,[341-43] some of the SU$_6$ deductions follow in fact already from the non-strange SU$_4$ algebra.

7.2. The Connections of the Static SU$_6$ Algebra with SU$_3$ × SU$_3$

We continue to study the model representation of Section 7.1 and assume for simplicity equal masses for the states. Then by Lorentz transformations we can relate matrix elements of currents and charges between the multiplet members at momentum $\bar{p}$ to their values in the rest system. Taking the common spatial momentum in the z direction and using invariant normalization: $\langle p|p' \rangle = (2\pi)^3 2E_p \delta^3(p-p')$, we relate

$$\langle a(\bar{p})|j_0|b(\bar{p}) \rangle = \langle a(\bar{0})|\frac{j_0+\beta j_z}{\sqrt{(1-\beta^2)}}|b(\bar{0}) \rangle, \quad \beta = \frac{p_z}{E} \qquad (7.5)$$

or written in a shorter notation:

$$_p\langle j_0 \rangle_p = {}_0\langle \frac{j_0+\beta j_z}{\sqrt{(1-\beta^2)}} \rangle_0; \quad _p\langle j_x \rangle_p = {}_0\langle j_x \rangle_0; \quad _p\langle j_z \rangle_p = {}_0\langle \frac{j_z+\beta j_0}{\sqrt{(1-\beta^2)}} \rangle_0$$
$$(7.6)$$

Equation (7.6) holds for vector and axial vector currents. We note that $_0\langle j_x \rangle_0 = 0, {}_0\langle j_z \rangle_0 = 0, {}_0\langle \bar{j}_0 \rangle_0 = 0$ and introduce unit normalization $\langle\langle p|p' \rangle\rangle = \delta^3(p-p')$. In the notation of (7.6) we find:

$$_p\langle\langle j_0 \rangle\rangle_p = {}_0\langle\langle j_0 \rangle\rangle_0, \qquad (7.7) \quad _p\langle\langle \bar{j}_0 \rangle\rangle_p = \beta\,{}_0\langle\langle \bar{j}_z \rangle\rangle_0. \quad (7.10)$$

$$_p\langle\langle \bar{j}_z \rangle\rangle_p = {}_0\langle\langle \bar{j}_z \rangle\rangle_0, \qquad (7.8) \quad _p\langle\langle j_z \rangle\rangle_p = \beta\,{}_0\langle\langle j_0 \rangle\rangle_0, \quad (7.11)$$

$$_p\langle\langle \bar{j}_x \rangle\rangle_p = \sqrt{(1-\beta^2)}\,{}_0\langle\langle \bar{j}_x \rangle\rangle_0, \quad (7.9) \quad _p\langle\langle j_x \rangle\rangle_p = 0. \qquad (7.12)$$

We see that the matrix elements of the generators $\bar{Q}^i_x$ and $\bar{Q}^i_y$ are suppressed by a factor $\sqrt{(1-\beta^2)}$. This implies that a multiplet of single-particle states cannot provide a representation of the algebra S[(7.7)–(7.9)] at all momenta simultaneously. If a multiplet provides a representation of the algebra at rest, multi-particle states have to supplement the saturation at $\bar{p} \neq 0$.[227] Only the collinear subalgebra, generated by Q^i_0 and $\bar{Q}^i_z$, is un-affected by the Lorentz transformation. It has the structure of $SU_3 \times SU_3$. There is nothing inconsistent in having a single-particle representation of the collinear algebra at all momenta $\bar{p} = (0,0,p_z)$, but this does not imply that the single-particle states $|s(\bar{p})\rangle$ are not connected to many-particle states $\langle m(\bar{p})|$ at all momenta $\bar{p}$. $\langle m(\bar{p})|\bar{Q}_z|s(\bar{p})\rangle \equiv 0$ would be sufficient to establish single-particle representations, but it is not necessary. Cole-man[106] has shown from basic axioms of quantum field theory that it is not possible to have $\langle m(\bar{p})|\bar{Q}_z|s(\bar{p})\rangle \equiv 0$ for all momenta $\bar{p}$ and all many-particle states $\langle m(\bar{p})|$ (see also Pohlmeyer[308]). But this is not at all required; there exists a model by Gerstein and Lee [170] demonstrating that many-particle contributions to commutators may add up to zero in the sum over intermediate states although not vanishing individually.

At infinite momentum we have the equality $_\infty\langle\langle \bar{J}_0\rangle\rangle_\infty = \,_\infty\langle\langle \bar{J}_z\rangle\rangle_\infty = \,_0\langle\langle \bar{J}_z\rangle\rangle_0$. The collinear subalgebra then becomes equivalent to the chiral $SU_3 \times SU_3$ algebra studied in Chapters 2–6, at least for the model under consideration. The representation of $SU_3 \times SU_3$ by the fifty-six baryon states at infinite momentum is reducible. Since the generators commute with rotations about the z-axis, they leave invariant the helicity of the states. So the representation space can be decomposed into four invariant subspaces of dimensions 10; 18; 18; 10, corresponding to helicities $\frac{3}{2}; \frac{1}{2}; -\frac{1}{2}; -\frac{3}{2}$. In $SU_3 \times SU_3$ the two commuting algebras can be pictured to operate with right-handed and left-handed quarks,[195] so one can immediately read off the $SU_3 \times SU_3$ content of the subspaces: (10,1); (6,3); (3,6); (1,10).†

Alternatively we can start by representing the chiral $SU_3 \times SU_3$ algebra in the model of fifty-six baryon states at infinite momentum and taking account of their helicity properties we can

---

† (a,b) means that of the two commuting SU_3 algebras the first one has a representation isomorphic to the a-dimensional one, the second one isomorphic to the b-dimensional one.

determine the matrix elements of the corresponding collinear subalgebra. By (7.7), (7.8) and (7.10) and by rotations in the rest frame we recover a representation of the full static SU_6 algebra S (in the limit of equal masses). It was found [references (23), (57), (169), (170), (194)] that a representation of $SU_3 \times SU_3$ in the model of fifty-six baryon states at infinite momentum already contains all the SU_6 results of Section 7.1. To avoid a misunderstanding we stress that here and in Section 7.1 only SU_6 results for axial and vector charges are discussed; for other SU_6 results, like the magnetic moment predictions, see Section 7.3.

The construction of a rest algebra from an arbitrary $SU_3 \times SU_3$ representation at infinite momentum is formally always possible, as long as the states are given degenerate masses, but we will obtain an SU_6 rest algebra only if several $SU_3 \times SU_3$ representations are combined according to the spins of the states. The derivation of chiral $SU_3 \times SU_3$ representations for states at infinite momentum from an SU_6 rest algebra has been demonstrated, but the reverse way, the introduction of angular momentum into representations of the $SU_3 \times SU_3$ current algebra at infinite momentum, is still under investigation.[114–115]†

Further complications arise if states with different masses are considered.[23] Then it is no longer possible to relate matrix elements of charges in different reference frames, because states of different masses will not preserve equal spatial momenta under Lorentz-transformations.

Furthermore, different kinematical forms in the currents will contribute to the charges. If a finite set of states will provide a representation of the algebra at all, then in general they will do so only at one momentum.

Finally we wish to sketch a partly dynamical approach to the SU_6 results of Section 7.1. Working in the approximation of SU_3 symmetry we want to find suitable dominant intermediate states for a commutator of two axial charges at infinite momentum (equivalently: zero momentum transfer) between octet baryons.

$$\sum_n \left[\frac{\langle a|\bar{Q}^i|n\rangle \langle n|\bar{Q}^j|b\rangle}{(2\pi)^3 2E_n} - \frac{\langle a|\bar{Q}^j|n\rangle \langle n|\bar{Q}^i|b\rangle}{(2\pi)^3 2E_n} \right]$$

$$= if^{ijk}\langle a|Q^k|b\rangle. \quad (7.13)$$

†See also (437), (510).

The baryon octet alone is not sufficient as there is no suitable eight-dimensional representation of $SU_3 \times SU_3$; thus higher states cannot be neglected. We have

$$\langle a|\bar{Q}^i|n\rangle \sim \langle a|\bar{j}_0^i|n\rangle \sim \frac{\langle a|\bar{D}^i|n\rangle}{(E_n-E_a)} \underset{p\to\infty}{\sim} \frac{\langle a|\bar{D}^i|n\rangle}{m_n^2-m_a^2} \approx \frac{f_M \cdot g_{Man}}{m_n^2-m_a^2}.$$

$$(7.14)$$

In the last step we used meson pole dominance (Chapter 5) for axial vector divergences: f_M is a meson decay constant, g_{Man} a generalized coupling constant. We see that important contributions will come from states which are:

(a) nearby in mass (small denominator in (7.14)),
(b) strongly coupled to baryons and pseudoscalar mesons (large numerator).

The obvious candidates are the lowest baryon–meson resonances. Keeping only the decuplet and neglecting all others gives the SU_6 results, but we know now how to improve the predictions: we just have to keep further intermediate states and use information about meson–baryon scattering.

The Adler–Weisberger relation is a good example. The $SU_3 \times SU_3$ model representation with degenerate baryons gives $|g_A| = 1\cdot67$. Taking into account mass splitting and finite N^*-width changes the predictions to $|g_A| \approx 1\cdot4$ (Section 5.2). Finally taking all contributions to the commutator we get $|g_A| \approx 1\cdot16$. This "dynamical approach" does not explain the $SU_3 \times SU_3$ algebra of course.

7.3. The Baryon Magnetic Moments and Mixing Schemes

The ratio $\mu_P^T/\mu_N^T = -\frac{3}{2}$ for the (total) nucleon magnetic moments has been one of the spectacular successes of the SU_6 symmetry model. Magnetic moments can only be measured if there is non-zero momentum transfer, so we have to consider commutators of the charges with the electromagnetic current if we want to deal with magnetic moments. In a natural generalization of (7.4) we assume for the static SU_6 algebra:

$$[Q_0^i, j_\mu^j(x)] = if^{ijk}j_\mu^k(x),$$

$$[\bar{Q}_\mu^i, j_0^j(x)] = if^{ijk}\bar{j}_\mu^k(x),$$

$$[\bar{Q}_l^i, j_m^j(x)] = ie_{lmn}d^{ijk}j_n^k(x) - i\delta_{lm}f^{ijk}\bar{j}_0^k(x) \, ; [l,m,n \neq 0]. \qquad (7.15)$$

110

In the limit of infinite momentum we have (2.16)–(2.19). In the rest frame we may form the first moment of the electromagnetic current

$$V_i^{\mathrm{el}} = e_{ijk} \int (dx)^3 x_j j_k^{\mathrm{el}}(x). \tag{7.16}$$

We find as a consequence of (7.15) that V_i^{el} transforms as a member of a *35*-plet under the SU$_6$ algebra S. With

$$\langle a(0) \,|\, V_i^{\mathrm{el}} \,|\, b(0) \rangle$$

$$= e_{ijk}\delta^3(\bar{q}) \left[(+i)\frac{\partial}{\partial q_j} \langle a(-\tfrac{1}{2}\bar{q}) \,|\, j_k^{\mathrm{el}}(0) \,|\, b(\tfrac{1}{2}\bar{q}) \rangle \right] \tag{7.17}$$

we can establish $\mu_P^T/\mu_N^T = -\tfrac{3}{2}$ for the total magnetic moments, because there is only one free parameter in the matrix elements of a *35*-plet operator between states of a *56* multiplet of SU$_6$. In the symmetry model it was necessary to make the additional assumption that the prediction refers to total, not anomalous magnetic moments. Here this problem is solved without ambiguities. The *35*-plet assignment for V_i^{el} in (7.16) follows from (7.15), and careful evaluation of (7.17) shows that the Sachs form factor with μ^T is projected.

These results have been given by Lee.[240] He went on to investigate the commutators of moments (7.16). The quark model of Section 2.3 allows the construction of local commutators of currents in equs (2.24)–(2.27). Assuming these local commutators to be true in physical reality one can deduce the commutators of moments. Their algebra will not close: the commutator of two first moments is a second moment $\sim \int (dx)^3 x_i x_j j_k(x)$, its matrix elements are expressed by charge radii and quadrupoles. The charges of the SU$_6$ algebra are scalars or vectors under rotations, the moments are higher tensors. Applied to states at rest, they produce transitions only to states of definite spins and parities. Lee[240] has attempted a saturation of certain commutators of moments, taken between nucleons at rest, by the decuplet resonances only and has deduced interesting results, for instance: $\mu_P^T = \tfrac{1}{6}\langle r_P^2 \rangle$ and $\langle r_N^2 \rangle = 0$ connecting nucleon magnetic moments and charge radii (see also Dashen and Gell-Mann[112–13]). A number of authors followed with different moment-commutators and different sets of resonances for their saturation between nucleons: references (58–61), (93), (127), (311), (313), (349).† The

† Recent contribution: (464).

experimental success is not uniform; for the details we refer to the original papers. Here we only want to make two general remarks:

(a) Extracting from the quark model local commutators implies of course much stronger dynamical assumptions than the postulate of charge-charge or charge-current commutators. In Chapter 8 we will investigate so-called Schwinger terms in local commutators, containing the gradient of a spatial δ-function

$$[j_0(x),j_e(y)]=A_e(x)\delta^3(x-y)+B(x)\frac{\partial}{\partial x_e}\delta^3(x-y). \quad (7.18)$$

In a spatial integration to construct charges some Schwinger terms vanish, others become irrelevant for the applications in Chapters 2–6, but if present at all, they will contribute in moment commutators.

(b) In direct sum rules at rest the intermediate states will contribute with positive momentum transfers $\Delta^2=(m_i^2-m_e^2)$. Non-resonant continua (about which we have no data) will be enhanced at such masses m_i where $\Delta^2=m_R^2$ reaches the mass of a resonance with the quantum numbers of a current.

Moment operators of the static SU_6 algebra do not have such simple connections to moment operators of the chiral $SU_3\times SU_3$ at infinite momentum as do the charges. Therefore we consider anew the matrix elements of the commutators of charges and currents (2.24)–(2.27) between baryons of momenta $\langle p_z+q_x|\ldots|p_z\rangle$ with $p_z\to\infty$. For the electromagnetic current between nucleons we have matrix elements of the type

$$\bar{u}(p+q)\underbrace{(F_1^{el}\cdot\gamma_\mu-F_2^{el}\cdot\sigma_{\mu\nu}q_\nu)}_{\text{from current}}\frac{u(p)\bar{u}(p)}{2E_p}\underbrace{\gamma_0(\gamma^5)u(p)}_{\text{from charge}} \quad (7.19)$$

In the limit $p_z\to\infty$ the additional momentum q_x in the final state does not change the form of the spinor, so the magnetic moment term in the current connects only states of different helicities. Under the $SU_3\times SU_3$ algebra the current transforms as a superposition $(1,8)+(8,1)$, as does its space integral which is in fact a generator. One can now prove that in the baryon model-representation: $(10,1)+(6,3)+(3,6)+(1,10)$ one cannot have an operator $(1,8)+(8,1)$ connecting different helicity states.

112

As an illustration consider a matrix element linking helicity $(\tfrac{3}{2})$ and $(\tfrac{1}{2})$ states:

$$\langle (10,1)|(1,8)+(8,1)|(6,3)\rangle = \langle (10,1)|(1,8)|(6,3)\rangle$$
$$+ \langle (10,1)|(8,1)|(6,3)\rangle$$
$$= \langle 10|1|6\rangle\langle 1|8|3\rangle + \langle 10|8|6\rangle\langle 1|1|3\rangle = 0 \qquad (7.20)$$

Because the two SU$_3$ algebras commute, matrix elements can be decomposed into products with the two factors referring to the two SU$_3$ algebras separately. All factors vanish due to the singlet parts. The same can be shown for a matrix element between helicity $(\tfrac{1}{2})$ and $(-\tfrac{1}{2})$ states:

$$\langle (6,3)\,|\,(1,8)+(8,1)\,|\,(3,6)\rangle$$
$$= \langle 6|1|3\rangle\langle 3|8|6\rangle + \langle 6|8|3\rangle\langle 3|1|6\rangle = 0. \qquad (7.21)$$

We see that the only solution is $F_2^{\mathrm{el}} = 0$. The anomalous magnetic moments of the baryon octet and the magnetic transition amplitudes in the octet–decuplet system have to vanish. This result, given by Dashen and Gell-Mann[114] and by Oehme,[284] is related to our comments about the determination of nucleon and hyperon magnetic moments in Section 5.4. A restriction to the SU$_2$ × SU$_2$ chiral algebra of strangeness conserving charges does not modify the result, because nucleon and N^* states would be assigned to a representation $(4,1)+(3,2)+(2,3)+(1,4)$, the electromagnetic current to $(3,1)+(1,3)$ ⟨isovector part⟩ and $(1,1)+(1,1)$ ⟨isoscalar part⟩. The arguments in (7.20) and (7.21) can still be applied to demonstrate $F_2^{\mathrm{el}} = 0$. It is unclear why the N^* dominance hypothesis which is in general inconsistent gives such good results in the particular sum rule of Section 5.4.

The least drastic modification necessitated by the failure in predicting the baryon magnetic moments would be to drop the decuplet dominance model and to keep further intermediate states in the commutator of axial charges taken between the members of the baryon octet.† These corrections are suggested by the success of the Adler–Weisberger relation.

A somewhat systematic inclusion of further intermediate states is described by the so-called mixing schemes. In the restricted Hilbert space of states at infinite momentum H_∞, we

---

†See in particular: (421).

can construct two sets of basis vectors: one for particle multiplets $|P_i\rangle$ and another for irreducible $SU_3 \times SU_3$ representations $|M_j\rangle$. The $SU_3 \times SU_3$ algebra has the mathematical property of compactness; this implies that we can decompose its (unitary) representation in H_∞ into finite dimensional irreducible ones, whose basis vectors are $|M_j\rangle$. In general the two bases need not coincide:

$$|M_j\rangle = \sum_i \alpha_{ji}|P_i\rangle \quad (7.22); \qquad |P_i\rangle = \sum_j \beta_{ij}|M_j\rangle \quad (7.23)$$

Equation (7.22) means that the charges of the algebra may connect different physical multiplets, and consequently by (7.23) a given multiplet of physical states has components in different representations of the algebra. As we see in the Adler–Weisberger relation the axial charges connect nucleons to an infinite number of physical multiplets: all states connected to the πP channel. The decuplet dominance hypothesis consisted in keeping only the $N^*(1236)$; in mixing schemes we keep further connections, but only a finite number. In deductions from equation (7.23) one assumes that the fifty-six physical baryon states have components in a fifty-six-dimensional representation of $SU_3 \times SU_3$ and in some others:

$$|56\rangle_P = \cos\vartheta\,|56\rangle_M + \sin\vartheta\,\cos\phi\,|X\rangle_M$$
$$+ \sin\vartheta\,\sin\phi\,\cos\chi\,|Y\rangle_M + \dots \quad (7.24)$$

Each new admixture introduces a new parameter which can be fitted to some experimental data and used to predict others. Several multiplets have been tentatively connected to the baryon fifty-six-multiplet in (7.24) but no clear picture has yet emerged. The large number of free parameters makes mixing schemes rather flexible, since higher resonances may be mixed even more extensively. We refer to the literature for details of the proposals: references (21–22), (62), (81), (89), (103), (159–62), (171–2), (191–3), (196), (243), (261), (273).†

A different interpretation of the difficulties with the baryon magnetic moments will be given in Chapter 9, in a classification of infinite momentum sum rules.

7.4. The U_{12} Current Algebra

The bilinear forms of the quark model contain more than an $SU_6 \times SU_6$ algebra; taking all Dirac matrices Γ in (7.4) we obtain

†Recent contributions: (391), (421), (435).

the algebra of (compact) U_{12}. The main difficulty in this extension is the fact that one does not know what physical meaning can be given to octets of tensor, scalar and pseudoscalar charges, whose commutators are determined by the algebra.

In the limit $p_z \to \infty$ (extensively reviewed by Dashen and Gell-Mann[114]) the tensor charges with components $(tx), (ty), (zx), (zy)$ can keep finite matrix elements; (zx) components become equivalent to (tx) components, (zy) to (ty). Their algebra can be associated with the collinear group $(SU_6)_p$. It is also possible to construct a dispersion theory for the tensor algebra[140] using

$$F_{\mu\gamma} = \int e^{iqx} \langle a | [T_{\mu\gamma}(x), h(0)] | b \rangle \vartheta(x_0)(dx)^4,$$

$$iq^\mu F_{\mu\gamma} + \int e^{iqx} \langle a | [\partial^\mu T_{\mu\gamma}(x), h(0)] | b \rangle \vartheta(x_0)(dx)^4$$

$$+ \int e^{iqx} \langle a | [T_{0\gamma}(x), h(0)] | b \rangle \delta(x_0)(dx)^4 = 0, \quad (7.25)$$

and taking the limit $q^\mu \to 0$. The interpretation of the tensor operators $T_{\mu\gamma}(x)$ has been attempted through proposed vector meson poles in its divergence $\partial^\mu T_{\mu\gamma}(x)$, but there are ambiguities in this process. The field is still under investigation, therefore we refer to the literature: references (4–5), (104), (116), (140), (225), (234), (283), (306), (353).†

Similarly one may try to interpret pseudoscalar currents indirectly by assuming that their form factors are dominated by pseudoscalar meson poles. Thus the pseudoscalar densities would provide an alternative set of interpolating fields for pseudoscalar mesons, which can be used in low-energy theorems.[229]‡ Sum rules can be evaluated by meson pole dominance approximations only if the momentum transfer between external and intermediate states is bounded; therefore we should take an infinite momentum limit. This causes difficulties for pseudoscalar charges: their matrix elements between states of fixed masses tend to zero:

$$\langle a(\bar{p}) | P(0) | b(\bar{p}) \rangle \approx \langle a(\bar{0}) | P(0) | b(\bar{0}) \rangle$$

$$\langle\langle a(\bar{p}) | P(0) | b(\bar{p}) \rangle\rangle \approx \frac{1}{\bar{p}} \langle\langle a(\bar{0}) | P(0) | b(\bar{0}) \rangle\rangle \to 0, \quad (7.26)$$

where we denote by $P(0)$ a pseudoscalar charge density. Disconnected configurations of infinite mass become dominant in

---

†Recent contributions: (391), (399), (425), (435), (467), (504), (512). ‡See also (462).

sum rules of pseudoscalar charges. There exists a prescription to reinterpret the limit,[297] but it introduces an additional assumption. As a result there are definite predictions made on the asymptotic behaviour of cross sections: references (266–7), (269). We refer to the literature for various applications of pseudoscalar densities both in low-energy theorems and sum rules: references (43), (220), (238), (248), (259), (266–7), (269).

In the quark model of Section 2.3 the scalar densities $(q^+(x)\gamma_0\frac{1}{2}\lambda_i q(x)) = s^i$ contain the mass operators of the quarks, both the average mass in s^0 and the mass splitting s^8. A possible interpretation of scalar densities along these lines has been indicated already by Gell-Mann;[166] details have been worked out in the following papers: references (29–30), (65), (74), (230), (340).

It has also been proposed that the scalar and pseudoscalar densities may occur in an effective Hamiltonian for non-leptonic decays. This assignment would satisfy the $\Delta I = \frac{1}{2}$ rule because the scalar and pseudoscalar densities are members of octets. In Section 6.5 we have seen how an effective $\Delta I = \frac{1}{2}$ rule emerges also from the (current) $\times$ (current) Hamiltonian; there is not sufficient evidence to abandon it.

It is tempting to assign the SU_3 breaking Hamiltonian $S^8 = \int s^8(x)\,(dx)^3$ and the effective Hamiltonian for parity conserving non-leptonic decays S^6 into the same octet of scalar densities. Such an assignment leads to substantial difficulties. In (3.8) we have shown that with $\mathcal{H}' = \lambda S^8$ follows:

$$i\frac{dQ^7}{dt} = i\int (dx)^3 \partial^\mu j_\mu^7(x) = [\mathcal{H},Q^7] = [\mathcal{H}',Q^7] = \lambda[S^8,Q^7]$$

$$= -i\frac{\sqrt{3}}{2}\lambda S^6. \tag{7.27}$$

This would mean that the parity conserving non-leptonic Hamiltonian density $s^6(x)$ $[S^6 = \int s^6(x)\,(dx)^3]$ is a current divergence and cannot produce a real transition:

$$\langle f|s^6(x)|i\rangle = -\frac{2}{\sqrt{3}\lambda}\langle f|\partial^\mu j_\mu^7(x)|i\rangle$$

$$= i(p^{\,f}-p^i)^\mu\left(-\frac{2}{\sqrt{3}\lambda}\right)\langle f|j_\mu^7(x)|i\rangle = 0, \tag{7.28}$$

by energy momentum conservation unless there is a pole in $\langle f|j^z_\mu(0)|i\rangle$. $j^z_\mu(0)$ connects only states of different strangeness which are never degenerate, so there is no dynamical reason to expect a pole. Furthermore, as far as form factor dispersion relations for vector currents are tested there is no reason to expect spurious singularities either. So we feel that the scheme of 'universal spurion coupling' has to be modified or abandoned. References to the work on non-leptonic decays with scalar and pseudoscalar densities are given subject to these reservations: references (1), (31), (37), (153), (158), (257–8), (338), (442), (447).

Finally an interesting suggestion has been made by Cabibbo, Horwitz and Ne'eman:[86,87] to construct an algebra from Regge pole vertex functions. It requires an extension of Regge theory to inelastic reactions before one can decide whether their proposal is a genuine advance or only a reformulation of a quark model.†

7.5. Other Extensions

Two kinds of further extensions should be noted:

(a) Use of spinor operators: Commutators between currents and charges with spinor operators or anticommutators between spinor operators are exploited. The difficulty with this work is that spinor operators are not observable; the proposals for their interpretation use the concepts of baryon fields. No final statement can yet be made about this work: references (45), (176), (199), (200), (249), (300), (430), (431), (471), (472).

(b) Use of current commutators for higher order electromagnetic effects like electromagnetic mass differences. Current commutators are proposed to be unchanged even by higher order electromagnetic effects. Following Fubini and Furlan[135] contributions have been made to the problem of electromagnetic mass differences: references (124), (250), (270), (361), (433); see also (481).

†See also recent suggestions of Okubo: (515), (516).

CHAPTER 8

Schwinger Terms

8.1. The Existence of Schwinger Terms

In extending the assumptions of charge–charge commutators (2.4), (2.7), (2.15) and charge–current commutators (2.16)–(2.19), we may try to use commutators of local currents. The quark model of Section 2.3 offers specific forms for such commutators in equations (2.24)–(2.27), for instance for vector currents

$$[j_0^i(x), j_\mu^j(y)]_{x_0 = y_0} = if^{ijk}j_\mu^k(x)\delta^3(x-y) \qquad (8.1)\,[(2.24)]$$

obtained by formal commutation of quark field operators in

$$j_\mu^i(x) = \bar{q}(x)\gamma_\mu\tfrac{1}{2}\lambda_i q(x). \qquad (8.2)\,[(2.21)]$$

The charge–current commutators can be derived from (8.1) by one spatial integration, charge–charge commutators by two. Operators integrated over space have non-vanishing matrix-elements only between states of equal spatial momentum; there is no such restriction for unintegrated operators. Consequently a commutator contains more information if it is not integrated.

On the other hand, the assumption of unintegrated commutators is much stronger; divergence terms for instance are lost in spatial integrations.

The full use of commutators like (8.1) has so far been inhibited by so-called Schwinger terms. It can easily be shown that the commutator (8.1) is not consistent.[351] We specialize (8.1) to $i=j$ and chose a hermitean and conserved current $j_\mu^i(x)$

$$[j_0^i(x), j_\mu^i(y)] = 0 \qquad\qquad \text{for } x_0 = y_0, \qquad (8.3)$$

$$[j_0^i(x), \sum_{e=1}^{3} \partial^e j_e^i(y)] = 0 \qquad\qquad \text{for } x_0 = y_0. \qquad (8.4)$$

By current conservation: $\sum\limits_{e=1}^{3} \partial^e j_e^i(y) = -\partial^0 j_0^i(y)$; specializing to $x = y$:

$$\left[j_0^i(0), \frac{\partial}{\partial x_0} j_0^i(0) \right] = -i[j_0^i(0),[H,j_0^i(0)]] = 0. \qquad (8.5)$$

We take the vacuum expectation value of (8.5)

$$0 = 2\langle 0|j_0^i(0)Hj_0^i(0)|0\rangle = 2\sum_n E_n|\langle 0|j_0^i(0)|n\rangle|^2. \qquad (8.6)$$

An inconsistency in (8.6) can only be avoided if $j_0^i(0)$ does not connect the vacuum to any particle-state,

$$\langle 0|j_\mu(x)|n\rangle = 0 \qquad (8.7)$$

as generalized by covariance. Operators satisfying (8.7) are unsuitable for describing physical currents, because they would not satisfy crossing. One can, in fact, rigorously show that local operators satisfying (8.7) must vanish identically.[S 26]

The source of this inconsistency is the inaccurate definition of local currents in (8.2) as the product of local operators. Defining the currents in the quark model as a limit with field operators at different space-time points in a time-ordered configuration[(351)]: $\lim\limits_{x'\to x} (\psi^+(x)\gamma^0\gamma_\mu\psi(x'))\epsilon(x-x')$ we still find (8.3) for commutators of two current-time components, but a different result for commutators of a current–time component with a current–space component:

$$[j_0^i(x),j_e^i(y)] = -iK\frac{\partial}{\partial x_e}\delta^3(x-y) \qquad (8.8)$$

where K is an infinite constant.

Difficulties in the construction of current commutators by formal commutation of the constituent field operators had already been found earlier in special cases [(177), (269), (310)]; they became generally known through the work of Schwinger.[(351)] Terms containing derivatives of δ-functions in commutators of local currents are now generally referred to as Schwinger terms.

In a spatial integration Schwinger terms are generally lost, particularly if their coefficients are constants, not operators. This kind of Schwinger terms is not dangerous; it would occur only in disconnected parts of matrix elements the effects of which can be easily eliminated.

In this first argument we used current conservation to demonstrate Schwinger terms; Okubo[298] demonstrated that they follow from more general requirements. His argument employs the Lehmann–Källen spectral representation, which asserts that the vacuum expectation value of a commutator formed with the components of a hermitean current can be represented as:

$$\langle 0|[j_\mu(x),j_\nu(y)]|0\rangle = \int\limits_{m_0}^{\infty} dm\left[-g_{\mu\nu}\rho_1(m) - \rho_2(m)\frac{\partial^2}{\partial x^\mu \partial x^\nu}\right]\Delta(x-y;m) \quad (8.9)$$

with positive spectral functions $\rho_1(m)$ and $\rho_2(m)$. For illustration we substitute a complete set of intermediate states:

$$\sum_n \left[\frac{\langle 0|j_\mu(0)|n\rangle\langle n|j_\nu(0)|0\rangle e^{ip_n(y-x)}}{(2\pi)^3 2E_n}\right.$$
$$\left. - \frac{\langle 0|j_\nu(0)|n\rangle\langle n|j_\mu(0)|0\rangle e^{ip_n(x-y)}}{(2\pi)^3 2E_n}\right]$$

$$= \int dm \int (dp)^4 \delta(p^2-m^2)[-g_{\mu\nu}\rho_1(m) + p_\mu p_\nu \rho_2(m)]\frac{[e^{ip(y-x)}-e^{ip(x-y)}]}{(2\pi)^3} \quad (8.10)$$

where

$$\sum_n \langle 0|j_\mu(0)|n(p)\rangle\langle n(p)|j_\nu(0)|0\rangle = -g_{\mu\nu}\rho_1(m)+p_\mu p_\nu \rho_2(m). \quad (8.11)$$

In (8.11) we take $\mu = \nu = 3$ and $p_3 = 0$. As a consequence of hermiticity we find $\rho_1(m) \geqslant 0$. We multiply (8.11) by $p^\mu p^\nu$, again as a consequence of hermiticity we find $m^2[-\rho_1(m) + m^2 \rho_2(m)] \geqslant 0$. We summarize the properties in $\rho_2(m) . m^2 \geqslant \rho_1(m) \geqslant 0$

$$\langle 0|[j_\mu(x),j_\nu(y)]|0\rangle = \int\limits_{m_0}^{\infty} dm\left[-g_{\mu\nu}\rho_1(m) - \rho_2(m)\frac{\partial^2}{\partial x^\mu \partial x^\nu}\right]$$

$$\int (dp)^4 \delta(p^2-m^2)\frac{(e^{ip(y-x)}-e^{ip(x-y)})}{(2\pi)^3}. \quad (8.12)$$

120

In the last integral we recognize the definition of $\Delta(x-y; m)$. This distribution has the properties: for $x_0 = y_0$, $\Delta(x-y; m) = 0$, $\partial/\partial x_0 \Delta(x-y; m) = -i\delta^3(x-y)$. We take now $\mu = 0$, $\nu \neq 0$ in (8.9). If the commutator were to vanish, (8.9) would give

$$0 = \langle 0|\,[j_0(x),j_\nu(y)]\,|0\rangle = i\,\frac{\partial}{\partial x^\nu}\,\delta^3(x-y)\int\limits_{m_0}^{\infty} \rho_2(m)\,dm.$$

$$(8.13)$$

It follows $\rho_2(m) = 0 \Rightarrow \rho_1(m) = 0 \Rightarrow j_\mu(x)|0\rangle = 0 \Rightarrow j_\mu(x) \equiv 0$ (excluding the unphysical case of massless hadrons). This argument is independent of current conservation and can also be applied for axial vector currents.

Little is known about the properties of Schwinger terms and most of the information is derived from special models. It is in the spirit of current algebras to assume only such classes of Schwinger terms whose existence is required either by general principles like covariance or by experimental evidence. The following is the present state of knowledge. There is no convincing demonstration (i.e. free of additional assumptions) that Schwinger terms have to be present in the commutator of current time-components. For commutators of space- with time-components the existence of Schwinger terms is certain, their operator character is disputed. For commutators of space- with space-components operator Schwinger terms should be expected.[83] A discussion of the detailed models and arguments is beyond the scope of this book, so we refer to the literature: references (13), (54), (83), (185), (186), (211), (239), (245).†

In the quark model Schwinger terms have special symmetry properties, being proportional to $\{\lambda^i\Gamma,\lambda^j\Gamma'\}_+$ in a commutator $[j_\Gamma^i(x),j_{\Gamma'}^j(y)]$; they can easily be eliminated by appropriate anti-symmetrization.[77,264] For example, with vector currents in $[j_0^i(x),j_\mu^j(y)]$, the combination

$$\tfrac{1}{2}([j_0^i(x),j_\mu^j(y)] - [j_0^j(x),j_\mu^i(y)]) = if^{ijk}j_\mu^k(x)\delta^3(x-y)$$

$$(8.14)$$

would be free from Schwinger terms. That the same is true for physical currents is of course an assumption.

---

†Recent contributions: (409), (411), (412), (413), (443), (446), (477), (485), (486).

8.2. Alternative Approaches to Schwinger Terms

Alternative approaches to Schwinger terms have been given by Björken[67] and Veltman.[375] It is difficult to be sure that these different approaches always deal with exactly the same phenomena though this can be shown in simple cases.

We consider again the integrals over retarded commutators investigated in Chapter 4:

$$T_{\mu\nu} = \int e^{iq_2 x} \langle p_2| [j_\mu^i(x), j_\nu^j(0)] |p_1\rangle \vartheta(x_0)\,(dx)^4, \tag{8.15}$$

$$U_\nu = \int e^{iq_2 x} \langle p_2| [D^i(x), j_\nu^j(0)] |p_1\rangle \vartheta(x_0)\,(dx)^4, \tag{8.16}$$

and perform the familiar partial integration

$$-iq_2^\mu T_{\mu\nu} = U_\nu + \int e^{iq_2 x}\delta(x_0)\,\langle p_2| [j_0^i(x), j_\nu^j(0)] |p_1\rangle. \tag{8.17}$$

We will no longer insist on the limit $q \to 0$ and therefore we may drop the construction of artificial mass-differences between external and intermediate states and we can consider the case $D^i(x) = 0 \Rightarrow U_\nu = 0$.

$$-iq_2^\mu T_{\mu\nu} = if^{ijk}\langle p_2| j_\nu^k(0)|p_1\rangle + (\text{Schwinger terms}). \tag{8.18}$$

If the current commutator is sufficiently regular at the origin, $T_{\mu\nu}$ will transform like a tensor and $q_2^\mu T_{\mu\nu}$ like a vector. For highly singular commutators the multiplication by $\vartheta(x_0)$ will produce non-covariant contributions; if these do not vanish upon multiplication by q_2^μ, then $q_2^\mu T_{\mu\nu}$ will not be a vector and there must also be non-covariant terms on the right-hand side of (8.18). If, for instance, the commutator $[j_0^i(x), j_\nu^j(y)]$ contains a term $K(\partial/\partial x^\nu)\delta^3(x-y)$ for $\nu \neq 0$, it would lead to the non-covariant term $-iKq_\nu(\nu \neq 0)$ in (8.18).

Retarded commutators or time-ordered products are not covariant operators in general, as has been recognized by Johnson.[210] The question is whether non-covariant parts appear only as constants in vacuum expectation values and disconnected parts of the amplitude $T_{\mu\nu}$, which can easily be eliminated, or whether they have operator character.

The association of Schwinger terms with non-covariant terms of a time-ordered product is due to Björken.[67] He also gave a

122

prescription for detecting these terms. Assume we start from the covariant amplitude $\tilde{T}_{\mu\nu}$ corresponding to $T_{\mu\nu}$, obtained say from lepton scattering in second-order perturbation theory. Both $\tilde{T}_{\mu\nu}$ and $T_{\mu\nu}$ are analytic in the complex q_0 plane apart from a cut along the real q_0 axis. They have the same discontinuity:

$$t_{\mu\nu} = \int e^{iq_2 x}\langle p_2|[j_\mu(x),j_\nu(0)]|p_1\rangle (dx)^4. \qquad (8.19)$$

Consequently $\tilde{T}_{\mu\nu}$ and $T_{\mu\nu}$ differ only by a polynomial in q_0. On the other hand, $T_{\mu\nu}$ can be integrated directly and gives (eq. (4.6)):

$$T_{\mu\nu} = \sum_n i\left\{ \frac{\langle p_2|j_\mu^i(0)|n\rangle\langle n|j_\nu^j(0)|p_1\rangle\delta^3(p_2+q_2-p_n)}{(2p_{n0})(q_{20}+p_{20}-p_{n0})} \right.$$

$$\left. - \frac{\langle p_2|j_\nu^j(0)|p_n\rangle\langle p_n|j_\mu^i(0)|p_1\rangle\delta^3(p_n+q_2-p_1)}{(2p_{n0})(q_{20}+p_{n0}-p_{10})} \right\}. \qquad (8.20)$$

If one is allowed to interchange the limit of the sum of intermediate states $\sum\limits_{n}$ with the limit $q_0 \to \infty$, then $T_{\mu\nu}(q_0)$ will tend to zero as $q_0 \to \infty$ with all other variables fixed. Björken's prescription is based on this observation. Explicitly, to obtain the retarded product he subtracts from the covariant amplitude $\tilde{T}_{\mu\nu}(q_0)$ all terms of order q_0^n with $n > 0$ in the limit $q_0 \to \infty$. The subtraction polynomial contains the Schwinger term. Unfortunately this limit is an unphysical one; if $\vec{q}$ is to remain fixed the limit would correspond to the scattering of particles with infinite masses: $m^2 = q_0^2 - \vec{q}^2$ in the scattering picture of Chapter 4. It is not clear whether the same arguments can be applied if q is allowed to vary along with q_0 to produce high energy limits of physical scattering processes. In any case it is possible[67] to give conditions on the asymptotic behaviour of $T_{\mu\nu}$ in its scalar variables like ν and q_2^2 to suggest the absence of Schwinger terms in the sense of Björken in connected parts of $T_{\mu\nu}$.

In principle there may also be gradient terms in the commutator

$$[j_0^i(x),j_\nu^j(y)] = if^{ijk}j_\nu^k(x)\delta^3(x-y) + 0^{ij}(x)\frac{\partial}{\partial x^\nu}\delta^3(x-y)$$

$$(8.21)$$

which would not obstruct the covariance of (8.18); so far

123

there is no clear evidence in favour of these terms for current commutators, but the situation may be different with current-divergence commutators.[245] They may be connected to subtraction constants.[427] Until we are forced to introduce them we will assume that such "covariant Schwinger terms" are absent.

If in (8.17) the current is not conserved, then the non-covariant contributions from U_ν have also to be considered. In applications to low-energy theorems Schwinger terms are often irrelevant because their contributions are presumably proportional to a vanishing momentum q. In sum rules they may affect the assumption of unsubtracted dispersion relations. There are many open problems with Schwinger terms; without making any specific criticism we would like to suggest that studies of the particularly singular phenomenon of Schwinger terms might require a higher standard of mathematical rigour than the usual framework of current algebra investigations can provide in the present experimental stage of the theory.

Note added in proof: The problem of Schwinger terms seems to be avoided in Veltman's formulation of current algebras[375] in terms of divergence conditions on weak and electromagnetic currents. These can be derived from current algebra[454] by taking the Hamiltonian (1.1) for electromagnetic and the W-boson picture (6.58) for weak interactions and assuming zero commutators for hadron currents with lepton field operators. Veltman's equations express the electromagnetic and weak contributions to the current divergences as products of lepton fields and hadron currents. Using model processes like W-boson scattering, Veltman derived important results of current algebras from his equations[375] without being affected by Schwinger terms.

An interesting interpretation of these results has been given through the formalism of gauge transformations by several authors [(55), (79), (304), (407), (410), (478), (479)], following earlier work by Schwinger (352). As it has been brought out most clearly by Bell[55], this approach will reproduce the results of commutators where at least one of the operators is a current time-component. This excludes some of the results of chapter 7 and some of Adler's sum rules in section 9.6, where commutators among current space-components are used[497]. This classification coincides with recent results of Dietz and Kupsch[428], based on the covariance properties of time-ordered products.

There have been only very few applications of Schwinger terms in predictions of experimental quantities (464), (465), (509); at the moment these are not enough to accept the existence of Schwinger terms other than required by general principles.

CHAPTER 9

The Dispersion Theory of Current Algebras, II

9.1. The Dispersion Theory of Local Current Commutators

In this chapter we will discuss the general dispersion theory of current algebras proposed by Fubini.[132] As in Chapter 4 we start with a partial integration, defining

$$T_\mu = \int e^{iq_2 x}\, \vartheta(x_0)\, \langle p_2| [j_\mu(x), h(0)] |p_1\rangle\, (dx)^4, \qquad (9.1)$$

$$U = \int e^{iq_2 x}\, \vartheta(x_0)\, \langle p_2| [D(x), h(0)] |p_1\rangle\, (dx)^4, \qquad (9.2)$$

$$iq_2^\mu T_\mu + U + \int e^{iq_2 x}\, \delta(x_0)\, \langle p_2| [j_0(x), h(0)] |p_1\rangle\, (dx)^4 = 0 \qquad (9.3)$$

Using now local commutators we need no longer take the limit

$q_2^\mu \to 0$ because if we postulate

$$\delta(x_0)\, [j_0(x), h(0)] = \delta^4(x)\, k(0) \qquad (9.4)$$

we can extract information from (9.3) with arbitrary q_2^μ:

$$iq_2^\mu T_\mu + U + \langle p_2| k(0) |p_1\rangle = 0. \qquad (9.5)$$

To illustrate the application of dispersion relations to (9.5) we choose the simplest possible example: $h(0)$ and $k(0)$ being scalar operators, the external particles spinless and of equal mass. Granted sufficient regularity of the commutators in (9.1) and (9.2), T_μ will be a vector, U will be a scalar.

This approach will be applicable to current components and divergences only with some precautions about Schwinger terms. We will assume that these are at most constants (not operators), so they contribute only if $\langle p_2|p_1\rangle \neq 0$. To eliminate them we will consider a limit with $\langle p_2|$ and $|p_1\rangle$ having slightly different momenta wherever an ambiguity may arise. If there are operator Schwinger terms in the commutators under consideration, then they will be separated from equations like (9.5) together with the noncovariant contributions to the integrals (9.1) and (9.2); the following arguments apply to the projection of the covariant part in equation (9.3). "Covariant Schwinger terms" have to be excluded by assumption.

We further define

$$t_\mu = \int e^{iq_2 x}\langle p_2|\,[\,j_\mu(x),h(0)\,]\,|p_1\rangle (dx)^4, \qquad (9.6)$$

$$u = \int e^{iq_2 x}\langle p_2|[D(x),h(0)]|p_1\rangle (dx)^4, \qquad (9.7)$$

$$iq_2^\mu t_\mu + u = 0 \quad \text{by partial integration.} \qquad (9.8)$$

We introduce the vector q_1 by $p_1 + q_1 = p_2 + q_2$ and $P_\mu = \frac{1}{2}(p_1 + p_2)_\mu$, $Q_\mu = \frac{1}{2}(q_1 + q_2)_\mu$, $\Delta_\mu = (p_2 - p_1)_\mu = (q_1 - q_2)_\mu$. As independent scalar variables we choose $\nu = P \cdot Q = P \cdot q_2 = Pq_1$, $t = (p_2 - p_1)^2$, q_1^2, q_2^2 as in Chapter 4.

We expand the tensors into independent kinematic forms:

$$T_\mu = A(\nu,t,q_1^2,q_2^2)P_\mu + B(\nu,t,q_1^2,q_2^2)Q_\mu + C(\nu,t,q_1^2,q_2^2)\Delta_\mu, \quad (9.9)$$

$$t_\mu = a(\nu,t,q_1^2,q_2^2)P_\mu + b(\nu,t,q_1^2,q_2^2)Q_\mu + c(\nu,t,q_1^2,q_2^2)\Delta_\mu, \quad (9.10)$$

$$U = U(\nu,t,q_1^2,q_2^2), \qquad (9.11)$$

$$u = u(\nu,t,q_1^2,q_2^2). \qquad (9.12)$$

The integrals (9.1) and (9.2) are retarded causal commutators. For a limited range of fixed t,q_1^2,q_2^2 they can be shown to be analytic in ν apart from a cut along the real ν-axis with their discontinuity given by (9.6) and (9.7). Specifically the coefficients of corresponding kinematic forms are related to each other, for example:

$$a(\nu,t,q_1^2,q_2^2) = \text{disc}_\nu A(\nu,t,q_1^2,q_2^2). \qquad (9.13)$$

We assume the validity of unsubtracted dispersion relations for the invariant matrix elements:

126

$$A(\nu,t,q_1^2,q_2^2) = \frac{1}{2\pi i} \int_{-\infty}^{+\infty} \frac{a(\nu',t,q_1^2,q_2^2)}{\nu'-\nu}d\nu'. \qquad (9.14)$$

In terms of invariants (9.5) and (9.8) read:

$$i[\nu A(\nu,t,q_1^2,q_2^2) + (q_2Q)B(\nu,t,q_1^2,q_2^2) + (q_2\Delta)C(\nu,t,q_1^2,q_2^2)]$$
$$+ U(\nu,t,q_1^2,q_2^2) + K(t) = 0, \qquad (9.15)$$

$$i[\nu a(\nu,t,q_1^2,q_2^2) + (q_2Q)b(\nu,t,q_1^2,q_2^2) + (q_2\Delta)c(\nu,t,q_1^2,q_2^2)]$$
$$+ u(\nu,t,q_1^2,q_2^2) = 0, \qquad (9.16)$$

where we have denoted $\langle p_2|k(0)|p_1\rangle$ by $K(t)$. Note that it does not depend on (ν,q_1^2,q_2^2). We substitute the Hilbert transforms (9.14) into (9.15) and obtain:

$$\frac{1}{2\pi i} \int \frac{d\nu'}{\nu'-\nu}\{i\nu a(\nu',t,q_1^2,q_2^2) + i(q_2Q)b(\nu',t,q_1^2,q_2^2)$$
$$+ i(q_2\Delta)c(\nu',t,q_1^2,q_2^2) + u(\nu',t,q_1^2,q_2^2)\} + K(t) = 0. \qquad (9.17)$$

We compare the bracket in (9.17) with (9.16) taken at $\nu = \nu'$ and find that the same expression appears apart from the factor ν of $a(\nu',t,q_1^2,q_2^2)$. (The quantities $(q_2Q)=\frac{1}{4}(3q_2^2+q_1^2-t);(q_2\Delta)=\frac{1}{2}(q_1^2-q_2^2-t)$ do not depend on ν.) We obtain:

$$\frac{1}{2\pi} \int \frac{d\nu'}{\nu'-\nu}(\nu a(\nu',t,q_1^2,q_2^2) - \nu'a(\nu',t,q_1^2,q_2^2)) + K(t) = 0, \qquad (9.18)$$

$$K(t) = \frac{1}{2\pi} \int d\nu' \cdot a(\nu',t,q_1^2,q_2^2). \qquad (9.19)$$

This is the generalized dispersion relation, valid for a continuous domain in the variables t,q_1^2,q_2^2. We may continue the integral (9.19) analytically in these parameters into its full analyticity domain. A number of interesting points arise in (9.19).

First we wish to establish the connection to the special case of Chapter 4: $q_2^2 = 0,q_1^2 = t$. Equation (9.16) then takes the form:

$$i\nu'a(\nu',t,t,0) + u(\nu',t,t,0) = 0, \qquad (9.20)$$

consequently:

$$K(t) = \frac{i}{2\pi} \int \frac{dv'}{v'} u(v',t,t,0).$$

(9.21)

This reproduces the special relation (4.10) of Fubini, Furlan and Rossetti,[136] if there are no contributions to $u(v',t,t,0)$ at $v'=0$. As we saw in Chapter 4, such contributions would come from degenerate intermediate states with the same mass as the external states. In Chapter 4 we explicitly excluded this case by introducing artificial mass differences δm in order to interpret $[u(v',t,t,0)/v']$ as a finite limit $\lim_{\delta m \to 0}[u(v',t,t,0)/v']$. The idea behind this treatment was a requirement of continuity between exact and slightly broken symmetries. We can now to a certain extent justify this procedure. The integrand $a(v',t,t,0)$ in (9.19) is not sensitive to small mass changes in intermediate states, it does not contain mass differences in a denominator, for instance. Thus with our limiting prescription of Chapter 4 we are in accordance with the general dispersion theory: $a(v',t,t,0) = \lim_{\delta m \to 0} a(v',t,t,0) = \lim_{\delta m \to 0} iu(v',t,t,0)/v'$.

We can now see a second point in this comparison. One of the achievements of the special dispersion relations of Chapter 4 was the fact that the sum rules required only a knowledge of the matrix elements of the current divergence, so they could be approximated by meson poles. For the more general dispersion relations here we need a knowledge of the current-matrix elements, if we want to go beyond (9.21). This is why these general sum rules can only be tested in electromagnetic reactions in the near future; there is no other current whose matrix elements are sufficiently well-known experimentally. For instance, a large number of sum rules has been worked out for photoproduction, electroproduction and Compton scattering; sum rules for neutrino reactions are similar but less likely to be tested. For the details we refer to the literature: references (11), (51), (66), (77), (82), (88), (102), (178–80), (182–3), (197–8), (232), (274). Here we will consider only the best known one in Section 9.3: the sum rule of Cabibbo and Radicati.[88] Its evaluation already reaches the limit of our experimental knowledge.

In Section 9.4 we will give some general considerations about the dependence of (9.19) on t and in particular on q_1^2 and q_2^2. Among other aspects this will lead us to the recently discussed super-convergence relations. But first we shall extend the general dispersion theory to a case more complicated than that of this section, namely where $h(0)$ is not a scalar but rather a vector or axial vector current. Conceptually nothing new will be added, only the formulas will become rather lengthy.

We also have to show the relations of the general dispersion theory to the infinite momentum approach.[144] This time this comparison will lead to some interesting classifications of sum rules.

The problem of convergence of the dispersion integral (9.19) and the validity of unsubtracted dispersion relations like (9.14) will be discussed in Section 9.5.

9.2. General Sum Rules for Spinless Particles

We will introduce the following tensors:

$$T_{\mu\nu} = \int e^{iq_2 x}\vartheta(x_0)\langle p_2|[j_\mu^i(x),j_\nu^j(0)]|p_1\rangle(dx)^4 \qquad (9.22)$$

$$U_\nu = \int e^{iq_2 x}\vartheta(x_0)\langle p_2|[D^i(x),j_\nu^j(0)]|p_1\rangle(dx)^4 \qquad (9.23)$$

$$U'_\mu = \int e^{iq_2 x}\vartheta(x_0)\langle p_2|[j_\mu^i(x),D^j(0)]|p_1\rangle(dx)^4 \qquad (9.24)$$

$$V = \int e^{iq_2 x}\vartheta(x_0)\langle p_2|[D^i(x),D^j(0)]|p_1\rangle(dx)^4. \qquad (9.25)$$

We also introduce the corresponding integrals $t_{\mu\nu}$, u_ν, u'_μ, v:

$$t_{\mu\nu} = \int e^{iq_2 x}\langle p_2|[j_\mu^i(x),D^j(0)]|p_1\rangle(dx)^4 \qquad (9.26)$$

and similarly the others. The momenta and scalar variables are the same as in Section 9.1. We will assume the commutators:

$$[j_0^i(x),j_\nu^j(0)] = if^{ijk}j_\nu^k(x)\delta^3(x), \qquad (9.27)$$

$$[j_0^i(x),D^j(0)] = c^{ijk}S^k(x)\delta^3(x) , \qquad (9.28)$$

where $S^k(x)$ is an unspecified scalar operator. Possible constant Schwinger terms have no effect on the deductions to follow and therefore they have been dropped. In (9.28) particularly the absence of operator Schwinger terms is a strong assumption, but most of the deductions will not depend on this equation.

We can perform two partial integrations

$$iq_2^\mu T_{\mu\nu} + U_\nu + if^{ijk}\langle p_2|j_\nu^k(0)|p_1\rangle = 0, \qquad (9.29)$$

$$iq_2^\mu U_\mu' + V + c^{ijk}\langle p_2|S^k(0)|p_1\rangle = 0, \qquad (9.30)$$

and another two by a translation:

$$T_{\mu\nu} = \int e^{iq_1x}\vartheta(x_0)\langle p_2|[j_\mu^i(0),j_\nu^j(-x)]|p_1\rangle(dx)^4 \qquad (9.31)$$

$$iq_1^\nu T_{\mu\nu} - U_\mu' + if^{ijk}\langle p_2|j_\mu^k(0)|p_1\rangle = 0, \qquad (9.32)$$

$$iq_1^\nu U_\nu - V - c^{jik}\langle p_2|S^k(0)p_1\rangle = 0. \qquad (9.33)$$

We combine these equations to give

$$q_2^\mu q_1^\nu T_{\mu\nu} - V - c^{jik}\langle p_2|S^k(0)|p_1\rangle + q_1^\nu f^{ijk}\langle p_2|j_\nu^k(0)|p_1\rangle = 0, \qquad (9.34)$$

$$q_2^\mu q_1^\nu T_{\mu\nu} - V - c^{ijk}\langle p_2|S^k(0)|p_1\rangle + q_2^\mu f^{ijk}\langle p_2|j_\mu^k(0)|p_1\rangle = 0, \qquad (9.35)$$

For the tensors $t_{\mu\nu}$, u_ν, u_μ' and v the same partial integrations give similar relations with the equal-time commutators omitted, for example:

$$iq_2^\mu t_{\mu\nu} + u_\nu = 0; \quad iq_2^\mu u_\mu' + v = 0. \qquad (9.36)$$

The next step is a decomposition into independent kinematic forms. To keep the problem manageable we will deal with spinless external particles only. The following arguments can also be applied to general particles when a spin average is taken. We choose the decomposition:

$$T_{\mu\nu} = AP_\mu P_\nu + B^{(1)}P_\mu Q_\nu + B^{(2)}P_\mu\Delta_\nu + B^{(3)}Q_\mu P_\nu + B^{(4)}\Delta_\mu P_\nu$$

$$+ C^{(1)}Q_\mu Q_\nu + C^{(2)}Q_\mu\Delta_\nu + C^{(3)}\Delta_\mu Q_\nu + C^{(4)}\Delta_\mu\Delta_\nu + C^{(5)}g_{\mu\nu}, \qquad (9.37)$$

$$t_{\mu\nu} = aP_\mu P_\nu + b^{(1)}P_\mu Q_\nu + \dots, \qquad (9.38)$$

$$U_\nu = LP_\nu + MQ_\nu + N\Delta_\nu \qquad (9.39)$$

$$U_\mu' = L'P_\mu + M'Q_\mu + N'\Delta_\mu, \qquad (9.40)$$

$$u_\nu = lP_\nu + mQ_\nu + n\Delta_\nu, \tag{9.41}$$

$$u'_\mu = l'P_\mu + m'Q_\mu + n'\Delta_\mu, \tag{9.42}$$

All invariant functions depend on the scalars ν, t, q_1^2, q_2^2. As in (9.15) and (9.16), we will now write the partial integrations (9.29), (9.30), (9.32), (9.33) and (9.36) in terms of invariants. When dealing with a vector equation, like (9.29), we can split it up into three independent parts, proportional to P_ν, Q_ν and Δ_ν:

$$i[(q_2 P)A + (q_2 Q)B^{(3)} + (q_2\Delta)B^{(4)}] + L + 2if^{ijk}F^+(t) = 0, \tag{9.43}$$

$$i[(q_2 P)B^{(1)} + (q_2 Q)C^{(1)} + (q_2\Delta)C^{(3)} + C^{(5)}] + M = 0, \tag{9.44}$$

$$i[(q_2 P)B^{(2)} + (q_2 Q)C^{(2)} + (q_2\Delta)C^{(4)} - \tfrac{1}{2}C^{(5)}]$$
$$+ N + if^{ijk}F^-(t) = 0, \tag{9.45}$$

using the notation: $\langle p_2 | j_\nu^k(0) | p_1 \rangle = 2F^+(t)P_\nu + F^-(t)\Delta_\nu.$ (9.46)

We now make the assumption that the invariant functions in (9.22)–(9.25) satisfy unsubtracted dispersion relations in ν as in equation (9.14). As in (9.17), we substitute the Hilbert transforms and use equation (9.36) for simplifying the expressions. As in (9.18), only terms containing a factor of ν will remain and give integrals equated to equal-time commutators (9.19). Explicitly, we obtain:

$$2if^{ijk}F^+(t) = \frac{1}{2\pi}\int d\nu'\,a(\nu',t,q_1^2,q_2^2) \qquad (9.47) \quad \text{from (9.29), (9.43),}$$

$$0 = \frac{1}{2\pi}\int d\nu'\,b^{(1)}(\nu',t,q_1^2,q_2^2) \quad (9.48) \quad \text{from (9.29), (9.44),}$$

$$if^{ijk}F^-(t) = \frac{1}{2\pi}\int d\nu'\,b^{(2)}(\nu',t,q_1^2,q_2^2) \quad (9.49) \quad \text{from (9.29), (9.45),}$$

$$2if^{ijk}F^+(t) = \frac{1}{2\pi}\int d\nu'\,a(\nu',t,q_1^2,q_2^2) \qquad\qquad \text{from (9.32),}$$

$$0 = \frac{1}{2\pi}\int d\nu'\,b^{(3)}(\nu',t,q_1^2,q_2^2) \quad (9.50) \quad \text{from (9.32),}$$

$$if^{ijk}F^-(t) = \frac{1}{2\pi} \int dv' \, b^{(4)}(v',t,q_1^2,q_2^2) \qquad (9.51) \quad \text{from (9.32)},$$

$$c^{ijk}\langle p_2|S^k(0)|p_1\rangle = \frac{1}{2\pi} \int dv' \, l'(v',t,q_1^2,q_2^2) \qquad (9.52) \quad \text{from (9.30)},$$

$$-c^{jik}\langle p_2|S^k(0)|p_1\rangle = \frac{1}{2\pi} \int dv' \, l(v',t,q_1^2,q_2^2) \qquad (9.53) \quad \text{from (9.33)}.$$

As in Section 9.1, we can again reproduce the results of the special dispersion theory of Fubini, Furlan and Rossetti[136] by fixing either $q_2^2 = 0$ and $q_1^2 = t$ or $q_1^2 = 0$ and $q_2^2 = t$ and by using (9.36) to introduce matrix elements of current divergences in the integrand. Since we are dealing with equal-mass particles $\langle p_2| \ldots |p_1\rangle$ we may take $q_1^2 = q_2^2 = t = 0$, and we find from (9.36)

$$iva(v,0,0,0) + l(v,0,0,0) = 0; \quad ivl(v,0,0,0) - v(v,0,0,0) = 0 \quad (9.54)$$

We may now transform (9.47) into the form of a generalized Adler–Weisberger relation:

$$2if^{ijk}F^+(t) = \frac{1}{2\pi} \int \frac{dv' \, v(v',0,0,0)}{(v')^2}. \qquad (9.55)$$

Even with spinless particles there is already considerable complexity in the kinematic forms. For external particles of spin $\frac{1}{2}$ there are thirty-two independent kinematic forms in $T_{\mu\nu}$ and each partial integration like (9.29) gives rise to eight sum rules. These have been worked out in full by Gourdin (179–80), (182) and by Muzinich.[274]†

Next we wish to consider the relations of the sum rules (9.47)–(9.51) to the infinite momentum approach. The arguments will be very similar to Section 4.2, so we will emphasize only the new features: the separation of kinematic forms. In $t_{\mu\nu}$ the momentum carried by the currents is $q_{2\mu}$ and $q_{1\nu}$:

$$t_{\mu\nu} = (2\pi)^4 \sum_n \{\langle p_2|j_\mu^i(0)|p_n\rangle_{q_2}\langle p_n|j_\nu^j(0)|p_1\rangle_{q_1}\delta^4(p_2+q_2-p_n)$$

$$-\langle p_2|j_\nu^j(0)|p_n\rangle_{q_1}\langle p_n|j_\mu^i(0)|p_1\rangle_{q_2}\delta^4(p_n+q_2-p_1)\}. \qquad (9.56)$$

†See also (470).

132

Thus we reproduce the correct momentum transfer if we consider the commutator

$$\left[\int e^{-i\vec{q_2}\cdot\vec{x}} j^i_\mu(x)(dx)^3, \int e^{-i\vec{q_1}\cdot\vec{y}} j^j_\nu(y)(dy)^3 \right]$$
$$= if^{ijk} \int e^{-i(\vec{q_1}+\vec{q_2})\cdot\vec{x}} j^k_\lambda(x)(dx)^3 \quad (9.57)$$

where $\lambda = \mu$ if $\nu = 0$ and $\lambda = \nu$ if $\mu = 0$.

We choose the z-direction for the infinite momentum. To have a useful separation of the kinematic forms we arrange q_1 and q_2 such that $Q = \frac{1}{2}(q_1 + q_2)$ has only an x-component and $\Delta = (q_1 - q_2) = (p_2 - p_1)$ has only a y-component. $P = \frac{1}{2}(p_1 + p_2)$ has diverging z and t-components which are asymptotically equal. Because all sum rules in (9.47)–(9.51) come from kinematic forms containing P_μ or P_ν in (9.38) we can always have either $\mu = 0$ or $\nu = 0$ in (9.57). All the sum rules are reproduced; (9.47) from the commutator of two time components, and (9.48)–(9.51) from the commutator of a space-component with a time-component. Because the energy differences between external and intermediate states tend to zero like $\Delta m^2/p_z$, there is a restriction on the parameters q_1^2, q_2^2 and t. Only if three-dimensional vectors $\bar{q}_1$ and $\bar{q}_2$ can be found such that $q_1^2 = -(\bar{q}_1)^2$, $q_2^2 = -(\bar{q}_2)^2$, $t = -(\bar{q}_1 + \bar{q}_2)^2$, can (9.47)–(9.51) be obtained as direct sum rules. But once we have obtained the sum rules in some range, this range may be extended by analytic continuation.

Nevertheless the sum rules (9.47) and (9.48)–(9.51) are not on the same footing. When the state normalization factors are taken into account which introduce a factor $\sim 1/(p_z)^2$, the terms proportional to $P_\mu P_\nu$ remain finite in the limit. Terms proportional to $P_\mu Q_\nu$ or $P_\mu \Delta_\nu$ would tend to zero, unless the commutator is scaled up by a factor P_z, for instance by omitting the normalization of the external states. This distinction will be discussed further in Section 9.5.

9.3. The Sum Rule of Cabibbo and Radicati

In this section we will derive the sum rule of Cabibbo and Radicati.[88] In the literature it turns up in different connections: (11), (51–52), (77), (82), (88), (114), (125), (139), (178–80), (183), (212), (221), (232), (274)†; we cannot display here the variations among these partly overlapping approaches. A possible

†See also (416).

access to the sum rule is the formalism of Section 9.2, in particular sum rule (9.47) applied to the commutator of isospin currents:

$$[j_0^1(x) + ij_0^2(x), j_0^1(y) - ij_0^2(y)] = [j_0^+(x), j_0^-(y)]$$
$$= 2j_0^3(x)\delta^3(x-y) \tag{9.58}$$

taken between protons (at rest) in the spin average.

In (9.47) we separate the integral into the neutron contribution and the continuum. With $p_1 = p_2 = p$ we have $q_1 = q_2 = q$, $P_\mu = p_\mu, Q_\mu = q_\mu, \Delta_\mu = 0$, only ν and q^2 are independent scalar variables. With the notation

$$\langle P(p)|j_\mu^+(0)|N(p+q)\rangle = \bar{u}_P(p)\,(F_1(q^2)\gamma_\mu$$
$$+ F_2(q^2)\sigma_{\mu\nu}q_\nu)u_N(p+q) \tag{9.59}$$

and the results of Section 4.2 the neutron contribution to $t_{\mu\nu}(\nu,q^2)$ takes the form

$$(2\pi)\delta\left(\nu - \frac{m_N^2 - m_P^2}{2}\right)\tfrac{1}{4}\mathrm{Tr}\{(\gamma p + m_P)\,(F_1(q^2)\gamma_\mu + F_2(q^2)\sigma_{\mu\lambda}q_\lambda)$$

$$(\gamma p + \gamma q + m_N)\,(F_1(q^2)\gamma_\nu - F_2(q^2)\sigma_{\nu\rho}q_\rho)\}$$

$$= (2\pi)\delta(\nu)2p_\mu p_\nu\{(F_1(q^2))^2 - q^2(F_2(q^2))^2\}$$

$$+ (\text{terms not proportional to } p_\mu p_\nu) \tag{9.60}$$

in the approximation $m_P = m_N$. So we have for (9.47)

$$2 = 2(F_1(q^2))^2 - 2q^2(F_2(q^2))^2 + \frac{1}{2\pi}\int\limits_{\mathrm{cont}} a(\nu',q^2)d\nu'. \tag{9.61}$$

At $q^2 = 0$ the sum rule reduces to the statement of vector current conservation: $F_1(0) = 1$, since for a conserved current $a(\nu',0)$ vanishes in the continuum. This can be shown by the equivalence of (9.61) to a direct sum rule from the commutator of isovector charges at infinite momentum. The derivative of (9.61) with respect to q^2 at $q^2 = 0$ gives the sum rule of Cabibbo and Radicati:[88]

$$0 = 4\left[\frac{\partial}{\partial q^2}F_1(q^2)\right]_{q^2=0} - 2[F_2(0)]^2 + \frac{1}{2\pi}\int\limits_{\mathrm{cont}}\left[\frac{\partial}{\partial q^2}a(\nu',q^2)\right]_{q^2=0}d\nu'. \tag{9.62}$$

In (9.62) the first term can be expressed in terms of the iso-vector charge radius:

$$F_1(t) = 1 + \tfrac{1}{6}\langle r^2\rangle_v \cdot t; \ \langle r^2\rangle_v = \langle r^2\rangle_P - \langle r^2\rangle_N. \qquad (9.63)$$

The second term in (9.62) contains the difference of the nucleon anomalous magnetic moments

$$F_2(0) = (\mu_P^A - \mu_N^A). \qquad (9.64)$$

Both (9.63) and (9.64) are a consequence of the CVC theory (Section 1.2). The continuum term in (9.62) can be expressed by photoproduction cross-sections. This can best be demonstrated by expanding

$$t_{\mu\nu}(\nu,q^2) = \int e^{iqx}\langle P|[j_\mu^+(x),j_\nu^-(0)]|P\rangle (dx)^4$$

$$= a(\nu,q^2)p_\mu p_\nu + b^{(1)}(\nu,q^2)p_\mu q_\nu + b^{(3)}(\nu,q^2)q_\mu p_\nu$$

$$+ c^{(1)}(\nu,q^2)q_\mu q_\nu + c^{(5)}(\nu,q^2)g_{\mu\nu} \qquad (9.65)$$

in the notation of (9.38). Due to vector current conservation we have

$$0 = q^\mu t_{\mu\nu}(\nu,q^2)$$

$$= (\nu a(\nu,q^2) + q^2 b^{(3)}(\nu,q^2))p_\nu$$

$$+ (\nu b^{(1)}(\nu,q^2) + q^2 c^{(1)}(\nu,q^2) + c^{(5)}(\nu,q^2))q_\nu, \qquad (9.66)$$

$$0 = q^\nu t_{\mu\nu}(\nu,q^2)$$

$$= (\nu a(\nu,q^2) + q^2 b^{(1)}(\nu,q^2))p_\mu$$

$$+ (\nu b^{(3)}(\nu,q^2) + q^2 c^{(1)}(\nu,q^2) + c^{(5)}(\nu,q^2))q_\mu. \qquad (9.67)$$

In (9.66) and (9.67) the brackets have to vanish individually; we derive

$$\nu\left[\frac{\partial}{\partial q^2}a(\nu,0)\right] + b^{(1)}(\nu,0) = 0; \ \ \nu b^{(1)}(\nu,0) + c^{(5)}(\nu,0) = 0. \qquad (9.68)$$

135

So the continuum integral in (9.62) can be rewritten into

$$\int_{\text{cont}} \left[\frac{\partial}{\partial q^2} a(v',0) \right] dv' = \int_{\text{cont}} \frac{c^{(5)}(v',0)}{(v')^2} dv' \qquad (9.69)$$

We isolate $c^{(5)}(v,q^2)$ in $t_{\mu\nu}(v,q^2)$, for instance by taking the external protons at rest, q in the z-direction and choosing $\mu = \nu = 1$.

$$-c^{(5)}(v,q^2) = (2\pi)^4 \sum_n \frac{\{\langle p|j_1^+(0)|n\rangle\langle n|j_1^-(0)|p\rangle\delta^4(p+q-p_n)}{(2\pi)^3 2E_n}$$

$$\frac{-\langle p|j_1^-(0)|n\rangle\langle n|j_1^+(0)|p\rangle\delta^4(p_n+q-p)\}.}{(2\pi)^3 2E_n} \qquad (9.70)$$

The isospin currents in (9.70) are hermitean adjoints of each other, so only the absolute squares of the matrix elements enter, summed over all intermediate states. One may now aim at writing (9.70) in the form of some total cross-section, but there are difficulties. Equation (9.70) is not related to a neutrino cross-section because it contains only the weak vector currents; one can relate it to photoproduction by "isovector photons", but then one has to separate intermediate states of isospin $(\frac{1}{2})$ and $(\frac{3}{2})$ which receive a different weight by the Clebsch–Gordan coefficients in (9.70). Collecting all terms in (9.62) we obtain the sum rule

$$\frac{\langle r^2 \rangle_v}{3} = (\mu_P^A - \mu_N^A)^2 + \frac{2}{\pi\alpha} \int \frac{2\sigma^{(1)}(s) - \sigma^{(3)}(s)}{(s - m_P^2)} ds. \qquad (9.71)$$

In (9.71) the quantities $\sigma^{(I)}(s)$ are total cross-sections for photoproduction of states with isospin $(I/2)$ from protons by isovector photons; α is the fine-structure constant, $s = m_n^2$ is the relativistic energy. At large s the cross-section difference will tend to zero to the extent that the scattering of isovector photons is connected by form factor dominance to the scattering of vector-mesons which are expected to satisfy a Pomeranchuk theorem: $\lim_{s\to\infty} [\sigma_{Pp}^+(s) - \sigma_{Pp}^-(s)] = 0$. The evaluation of the integral is difficult. Unless one knows the isospin of the final states, for instance with resonances, it is not possible to separate the isospin parts in (9.71) with the measured photoproduction cross-sections; we would require a knowledge of the amplitudes. The

136

nucleon terms in (9.71) already balance each other approximately, the N^* (1236) contribution would destroy the agreement, but terms from higher $I = \frac{1}{2}$ resonances and low-energy background contributions may restore it.[175]

A similar sum rule can be derived for the pion charge radius: (88), (102), (175), (232). In this and all the other photoproduction, electroproduction and Compton scattering sum rules, referred to in Section 9.1, there are similar difficulties for accurate tests as in the sum rule of Cabibbo and Radicati.[88]

9.4. Sum Rules for Finite Momentum Transfer, Superconvergence Relations

In this section we consider in detail the dependence of the sum rules (9.47)–(9.51) on the variables t, q_1^2 and q_2^2. The integrands are the coefficients of certain kinematic forms in

$$t_{\mu\nu}(\nu,t,q_1^2,q_2^2) = (2\pi)^4 \sum_n \left\{ \frac{\langle p_2|j_\mu^i(0)|n\rangle\langle n|j_\nu^j(0)|p_1\rangle\delta^4(p_2+q_2-p_n)}{(2\pi)^3 2E_n} \right.$$
$$\xleftarrow{\quad} q_2 \xrightarrow{\quad} \xleftarrow{\quad} q_1 \xrightarrow{\quad}$$

$$\left. - \frac{\langle p_2|j_\nu^j(0)|n\rangle\langle n|j_\mu^i(0)|p_1\rangle\delta^4(p_n+q_2-q_1)}{(2\pi)^3 2E_n} \right\}.$$
$$\xleftarrow{\quad} q_1 \xrightarrow{\quad} \xleftarrow{\quad} q_2 \xrightarrow{\quad} \qquad (9.72)$$

In the current matrix elements the momentum transfer is given by q_1^2 and q_2^2; a t-dependence in the right-hand side can only arise through scalar products of $(q_1 q_2)$ or $(p_1 p_2)$ in the polarization sums of intermediate states. Intermediate states of finite spin produce only finite polynomials in t. Thus the generalized dispersion relations can never be saturated with a finite number of multiplets, because the current form factors on the left-hand sides of (9.47)–(9.51) have singularities for positive values of t. Such singularities can only be generated if intermediate states of arbitrarily high spin contribute; that is singularities in $F(t)$ involve the excitation of all partial waves in the crossed channel.

Approximate saturation of sum rules from local current commutators by a finite set of particle and resonance multiplets has been studied already quite early by Balachandran, Kummer and Pietschmann[44],[63],[236] with the commutator of isospin currents. For the full $SU_3 \times SU_3$ algebra it has been studied in particular by Oehme[285-7] (see also Section 7.3) in the infinite-momentum approach (see eq. (9.57)). We refer to his work for details; here we note only a general trend. Considering the case $q_1 \neq 0$ and $q_2 \neq 0$ in (9.57) requires a larger multiplet for approximate saturation of the commutators than the case $q_1 = 0$ or $q_2 = 0$. This is well illustrated with the SU_3 algebra, considering (9.57) between octet baryons. To obtain non-trivial form factors in certain lowest orders of q_1 and q_2, the baryon decuplet is needed as intermediate states.[286] (The results are related to the collinear U_6 symmetry model.) It is then not astonishing to find that commutators of the larger $SU_3 \times SU_3$ algebra taken between octet baryons would require an even larger set of intermediate states to give non-trivial results in the same lowest orders of q_1 and q_2. For general considerations of related problems see references (48) and (378).†

The independence of the integrals in (9.47)–(9.51) of q_1^2 and q_2^2 is even more interesting. Singularities of the integrand in q_1^2 and q_2^2 have to balance each other. We assume vector meson poles in the matrix elements of the currents, for simplicity at equal masses: $q_1^2 = \mu^2$, $q_2^2 = \mu^2$. We can then multiply for instance (9.47) by $(q_1^2 - \mu^2) \cdot (q_2^2 - \mu^2)$, continue analytically to the poles and obtain for the residue $\hat{a}(\nu', t, \mu^2, \mu^2)$

$$0 = \int d\nu' \hat{a}(\nu',t,\mu^2,\mu^2). \tag{9.73}$$

The residue $\hat{a}(\nu',t,\mu^2,\mu^2)$ is proportional to the ν-discontinuity of an invariant matrix element in the scattering amplitude of spin zero and spin one mesons:

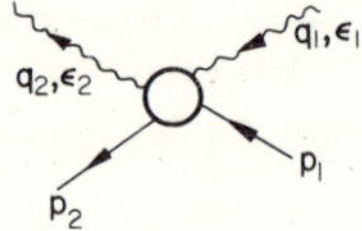

$$T = (\epsilon_1 P)(\epsilon_2 P)\hat{A}(\nu,t) + \tfrac{1}{2}\{(\epsilon_1 P) \cdot (\epsilon_2 Q) + (\epsilon_2 P)(\epsilon_1 Q)\} \hat{B}(\nu,t)$$
$$+ (\epsilon_1 Q)(\epsilon_2 Q)\hat{C}(\nu,t) + (\epsilon_1 \cdot \epsilon_2)\hat{D}(\nu,t). \tag{9.74}$$

Relation (9.73) is independent of the current commutator, because the left-hand side of (9.47) tends to zero, when multiplied by $(q_1^2 - \mu^2)(q_2^2 - \mu^2)$, whatever its value. We can even allow for various kinds of covariant Schwinger terms, so long as they do

†See also (403), (432).

not have poles at $q_1^2 = \mu^2$ and $q_2^2 = \mu^2$. Since the relation is independent of the commutator one would expect it to follow from more general principles than current algebras.

De Alfaro, Fubine, Furlan and Rossetti[19] identified (9.73) as a "super-convergence relation". If $\hat{A}(\nu,t)$ satisfies an unsubtracted dispersion relation in the energy variable ν and if for large ν: $|\hat{A}(\nu,t)| < c \cdot \nu^{-1-\epsilon}$ then we have

$$0 = \lim_{\nu \to \infty} \nu \hat{A}(\nu,t) = \lim_{\nu \to \infty} \frac{1}{2\pi i} \int \frac{\mathrm{disc}\,\hat{A}(\nu',t)}{(1 - \nu'/\nu)} d\nu'$$

$$= \frac{1}{2\pi i} \int \mathrm{disc}\,\hat{A}(\nu',t)\, d\nu' = 0 \qquad (9.75)$$

The basis of this superconvergence relation can be found in theorems on the asymptotic behaviour of scattering amplitudes at large energies. As a consequence of analyticity and unitarity, scattering amplitudes of spinless particles are asymptotically bounded by the Froissart limit

$$\hat{A}(\nu) \sim \nu \cdot (\log \nu)^2 \qquad (9.76)$$

which, disregarding logarithmic factors, leads to asymptotically constant total cross-sections.

Asymptotically constant total cross-sections are also found in the scattering of particles with spin; spin-flip reactions seem to be suppressed at high energies. In calculating total cross-sections from matrix elements like (9.74) additional powers of ν appear in the polarization sums over the vector mesons. They can be balanced to produce constant cross-sections only if the invariants $\hat{A}, \hat{B}, \hat{C}, \hat{D}$ are more strongly bounded than in the spinless case. Detailed analysis, again disregarding logarithmic factors, shows for (9.74).

$$|\hat{A}(\nu,0)| \leqslant \mathrm{const.}\, \nu^{-1}, \qquad (9.77)$$

$$|\hat{B}(\nu,0)| \leqslant \mathrm{const}, \qquad (9.78)$$

$$|\hat{C}(\nu,0)| \leqslant \mathrm{const.}\, \nu, \qquad (9.79)$$

$$|\hat{D}(\nu,0)| \leqslant \mathrm{const.}\, \nu. \qquad (9.80)$$

Equation (9.77) is not quite sufficient for deriving superconvergence relations, but assuming Regge pole[S 28] behaviour these bounds are not always reached, particularly if large internal quantum numbers are exchanged. By antisymmetrizing in the isospin or SU_3 indices one can always eliminate the Pomeranchuk trajectory.

One can now select amplitudes which are known to be strongly damped at high energies or take suitable linear combinations to subtract dominant trajectories and try to form sum rules with their superconvergence relations by keeping only a limited number of intermediate states in integrals like (9.73). The first successes of this approach have been overemphasized, but nevertheless useful sum rules emerge if a sufficiently large number of intermediate states is kept: references (19), (108), (131), (281), (289).

One may further conjecture superconvergence for particular transition amplitudes not related to commutators and produce sum rules by substituting selected intermediate states. Several results of current algebras, for instance the sum rule of Cabibbo and Radicati, have been reproduced in this approach by research groups at Dubna: references (32), (125), (212), (262–3), (356). This work is very promising, but it is still in a preliminary state because the authors have not yet given sufficiently general principles for which classes of amplitudes superconvergence relations should be postulated.

The derivation of superconvergence relations by equating pole residues in (9.73) is related in spirit to the extraction of singularities proposed by Jacob.[208–9]†

9.5. On the Asymptotic Behaviour of Current Commutator Matrix Elements

The Froissart bound does not permit the deduction of superconvergence relations for the scattering of spinless particles. It is comforting to see that current algebras do not give them either. We might, for instance, take axial vector currents in (9.22) and continue to the pseudoscalar meson poles in q_1^2 and q_2^2. We find, however, that none of the integrands in (9.47)–(9.53) contain these poles. According to the factorization of pole residues $t_{\mu\nu}$ is proportional to

†Recent contributions to Superconvergence relations: (394), (395), (401), (402), (448), (455), (457), (468), (476), (480), (483), (484), (494), (506), (507).

$$t_{\mu\nu} \sim \sum \left\{ \frac{\langle 0|\bar{j}^i_\mu(0)|\pi(q_2)\rangle\langle\pi(q'_2),p_2|t|n\rangle}{(m^2_\pi - q^2_2)} \right.$$

$$\cdot \frac{\langle n|t|p_1\pi(q'_1)\rangle\langle\pi(q_1)|\bar{j}^j_\nu|0\rangle}{(m^2_\pi - q^2_1)}\delta^4(p_2 + q_2 - p_n)$$

$$- \frac{\langle p_2|t|n\pi(q'_1)\rangle\langle\pi(q_1)|\bar{j}^j_\nu(0)|0\rangle}{(m^2_\pi - q^2_1)}$$

$$\left. \cdot \frac{\langle 0|\bar{j}^i_\mu(0)|\pi(q_2)\rangle\langle\pi(q'_2)n|t|p_1\rangle}{(m^2_\pi - q^2_2)}\delta^4(p_n + q_2 - p_1) \right\}$$

$$\sim q_{1\nu}q_{2\mu}. \tag{9.81}$$

With $q_{1\mu} = Q_\mu + \frac{1}{2}\Delta_\mu$ and $q_{2\mu} = Q_\mu - \frac{1}{2}\Delta_\mu$ we see that contributions with the pseudoscalar pole always occur in kinematic forms for which no sum rules are deduced. [This does not invalidate the application of pion pole dominance in cases like the Adler–Weisberger relation; by equations like (9.54) the discontinuities $a(\nu,t,0,0)$ of the dispersion integrals are related to discontinuities $v(\nu,t,0,0)$ which have a pion pole. This simple connection (9.55) is lost for $q^2_1 \neq 0, q^2_2 \neq 0$.]

On the other hand it is somewhat disturbing that the arguments of Section 9.4, as illustrated by (9.47), could equally well have been applied to (9.48)–(9.53). We would obtain a superconvergence relation for $B(\nu,t)$ in (9.74) and for the residues of $L(\nu,t, q^2_1,q^2_2)$ at $q^2_1 = \mu^2$ and of $L'(\nu,t,q^2_1,q^2_2)$ at $q^2_2 = \mu^2$ (equations (9.39) and (9.40)), which, as opposed to the case of $\hat{A}(\nu,t)$, are not supported by analyticity and unitarity arguments.

Some of the integrals in (9.48)–(9.53) vanish by applying crossing symmetry; this is not always the case. We will see later that there is good reason to doubt the correctness of the superconvergence relations derived from (9.48)–(9.53), unless there is a dynamical suppression, for instance associated with non-octet exchange. Therefore we must examine which assumptions might have been wrong in their derivation.

We propose to examine the assumption of unsubtracted dispersion relations for all amplitudes of $T_{\mu\nu}$, U_μ, U'_ν, V in equations (9.22)–(9.25). For this purpose we write out equations (9.43)–(9.44) and related equations for the simplified case: $p_1 = p_2, q_1 = q_2$:

$$i[\nu A(\nu,q^2) + q^2 B^{(3)}(\nu,q^2)] + L(\nu,q^2) + 2if^{ijk}F^+(0) = 0$$

$$(9.82) \ [(9.43)],$$

$$i[\nu B^{(1)}(\nu,q^2) + q^2 C^{(1)}(\nu,q^2) + C^{(5)}(\nu,q^2)] + M(\nu,q^2) = 0$$

$$(9.83) \ [(9.44)],$$

$$i[\nu L(\nu,q^2) + q^2 M(\nu,q^2)] - V(\nu,q^2) - c^{jik}\langle p_2|S^k(0)|p_1\rangle = 0$$

$$(9.84) \ [(9.33)].$$

In using unsubtracted dispersion relations it was assumed that all amplitudes vanish in the limit $\nu \to \infty$, and the sum rules (9.47), (9.48) and (9.53) may be read:

$$\lim_{\nu \to \infty} i(\nu A(\nu,q^2)) = -2if^{ijk}F^+(0), \qquad (9.85)$$

$$\lim_{\nu \to \infty} i(\nu B^{(1)}(\nu,q^2)) = 0, \qquad (9.86)$$

$$\lim_{\nu \to \infty} i(\nu L(\nu,q^2)) = c^{jik}\langle p_2|S^k(0)|p_1\rangle. \qquad (9.87)$$

Thus we have a new form of representing the equal-time commutator: as a high-energy limit of a scattering amplitude off the mass shell. Examining the residues of vector meson poles at $q^2 = \mu^2$ in $A(\nu,q^2)$, etc., in these equations we obtain the superconvergence relations directly. This connection of equal-time commutators with high energy limits has been only gradually understood; one of the first authors using this idea was Gasiorowicz;[154–5] but see also references (245) and (280).†

On the other hand it is simply not true that $\lim V(\nu,q^2) = 0$, as can be seen with the axial vector currents of the Adler–Weisberger relation. Taking pions as external particles we have for small q^2:

$$V_{\pi\pi}(\nu,q^2) = \int e^{iqx}\vartheta(x_0)\langle \pi^+(p)|[\bar{D}^+(x),\bar{D}^-(0)]|\pi^+(p)\rangle (dx)^4$$

$$\approx \frac{-(f_\pi \cdot m_\pi^2)^2\langle \pi^+(p)\pi^-(q')|t|\pi^+(p)\pi^-(q')\rangle}{(m_\pi^2 - q^2)^2} \qquad (9.88)$$

using pion pole dominance. If our present ideas are correct, the

---

†Such high energy limits are suitable to investigate sum rules in perturbation theory: (443), (444).

142

pion scattering amplitude in the pole residue is dominated by Pomeranchon exchange and tends to

$$\langle \pi^+(p),\pi^-(q')|t|\pi^+(p),\pi^-(q')\rangle \sim \nu \text{ const as } \nu \to \infty \quad (9.89)$$

up to logarithmic factors whose presence is not established. For protons in the external states the asymptotically linear energy dependence of the scattering amplitude is confirmed by asymptotically constant (πP) total cross-sections.

Thus, unless there is something fundamentally wrong with pion pole dominance at high energies, $V_{\pi\pi}(\nu,q^2)$ does not tend to zero. Its isospin anti-symmetric part is dominated by the ρ-trajectory and does not tend to zero either. Furthermore, $V_{\pi\pi}(\nu,q^2)$ does not satisfy an unsubtracted dispersion relation: $\int d\nu'(v_{\pi\pi}(\nu',0)/\nu')$ diverges. This can be most easily demonstrated if we take protons as external states, averaged over spin; we obtain (by the same method as in the Adler–Weisberger relation)

$$\int \frac{d\nu' v_{\pi P}(\nu',0)}{\nu'} = |f_\pi|^2 \int\limits_{(m_P+m_\pi)^2}^{\infty} \frac{m_P ds}{(s-m_P^2)}\left[\sigma^{\text{tot}}_{P\pi-}(s) + \sigma^{\text{tot}}_{P\pi+}(s)\right]\frac{p_\pi^L}{\pi} \to \infty. \quad (9.90)$$

This means that we have to give up dispersion sum rules based on commutators of currents and divergences (equations (9.52), (9.53), (9,87)) except in special cases where the high-energy amplitude is suppressed dynamically by the exchange of large internal quantum numbers. These cases include the derivation of the Gell-Mann Okubo mass formula in Section 4.2, an estimate of the (D/F) ratio of the axial vector current in the baryon octet[16,355] and an estimate of the σ-meson parameters.[152] In none of these cases can octet states be exchanged in the t-channel, because either isospin two or strangeness two is required. It is not quite clear whether applications to pion photoproduction can be maintained.[155,271]

Having dropped (9.52), (9.53) and (9.77) we must still proceed further. In (9.84) $M(\nu,q^2)$ and $V(\nu,q^2)$ have a double pseudoscalar meson pole at $q^2 = m_\pi^2$ (when dealing with axial vector currents); $L(\nu,q^2)$ has only a single pseudoscalar pole. Consequently from (9.88) $M(\nu,q^2)$ cannot tend to zero as ν tends to

143

infinity, because the residue of $V(\nu,q^2)$ does not tend to zero. Thus we lose the basis of equations (9.48), (9.50) and (9.53).

With arguments like this all the sum rules (9.48)–(9.53) may be invalidated in general; research about the details is still in progress. Similarly the high energy limits (9.86) and (9.87) should be abandoned; only sum rule (9.47) and the high-energy limit (9.85) survive these difficulties:

$$2if^{ijk}F^+(t) = \frac{1}{2\pi} \int d\nu'\, a(\nu',t,q_1^2,q_2^2)$$

$$= \lim_{\nu \to \infty}(-i\nu)\, A(\nu,t,q_1^2,q_2^2), \qquad (9.91)$$

where the integrand should be antisymmetrized in isospin indices if an exclusion of the Pomeranchuk trajectory is necessary to avoid ambiguities.

It is remarkable that (9.91) is the sum rule distinguished in the infinite-momentum approach in Section 9.2. Summing up the momentum-limits at individual intermediate states with fixed masses involves the neglect of disconnected configurations, as shown in Section 3.2. This omission may well be possible as long as the commutator is not scaled up by a factor P_z, as was necessary in the derivation of the sum rules (9.48)–(9.52) in Section 9.2. If the contributions of disconnected states vanish like $[(\text{const})/P_z]$ they will not appear normally, only when the commutator is scaled up. In the simple case of a commutator between a particle state and the vacuum the inconsistency of unsubtracted dispersion relations for all amplitudes in $T_{\mu\nu}$ has been demonstrated explicitly[333] (See also (371).)

In this connection we remark that the sum rules of Section 7.3 which demand the vanishing of the baryon anomalous magnetic moments in the decuplet dominance limit are also of this disputable class. At present we cannot decide whether these sum rules in Section 7.3 are valid when taking into account higher states or whether they should be abandoned; we shall reserve our judgement.

General considerations about the validity of unsubtracted dispersion relations have been given on the basis of Jost–Lehmann–Dyson representations;[174],[350] Feynman graphs,[45] Regge-pole theory.[134],[274] The subject of superconvergence relations will play an important role in a forthcoming book on current algebras by Fubini, Furlan and Rossetti (*Current Algebras*; North-Holland Publishing Company, Amsterdam).†

9.6. Other Classes of Dispersion Sum Rules

In this section we want to deal briefly with three further classes of dispersion sum rules. The first class contains Adler's sum rules, formulated as tests of current algebras in high-energy neutrino reactions;[11],[14],[66] in the framework of dispersion theory they can be derived as follows:

$$T_{\mu\nu} = \int e^{i(q_1+q_2)x/2}\vartheta(x_0)\langle p_2|[j_\mu^i(x/2),j_\nu^j(-x/2)]|p_1\rangle(dx)^4. \quad [(9.22)]$$

We perform a partial integration:

$$0 = i/2(q_1+q_2)^\lambda T_{\mu\nu} + \int e^{i(q_1+q_2)x/2}\left(\frac{\partial}{\partial x_\lambda}\vartheta(x_0)\right)\langle p_2|[j_\mu^i(x/2),$$

$$j_\nu^j(-x/2)]|p_1\rangle(dx)^4 + \int e^{i(q_1+q_2)x/2}\vartheta(x_0)\left\{\langle p_2|\left[\frac{\partial}{\partial x_\lambda}j_\mu^i(x/2),j_\nu^j(-x/2)\right]\right.$$

$$\left. + \left[j_\mu^i(x/2),\frac{\partial}{\partial x_\lambda}j_\nu^j(-x/2)\right]|p_1\rangle\right\}(dx)^4. \quad (9.92)$$

Equation (9.92) contains an arbitrary direction λ. It must be time-like to produce an equal-time commutator in the third term, so we may choose the direction of P^λ and obtain

$$0 = i\nu T_{\mu\nu} + \int e^{iq_2 x}\left(P_\lambda\frac{\partial}{\partial x_\lambda}\vartheta(x_0)\right)\langle p_2|[j_\mu^i(x),j_\nu^j(0)]|p_1\rangle$$

$$+ \tfrac{1}{2}\int e^{iq_2 x}\vartheta(x_0)\{\langle p_2|[P_\lambda\partial^\lambda j_\mu^i(x),j_\nu^j(0)]$$

$$- [j_\mu^i(x),P_\lambda\partial^\lambda j_\nu^j(0)]|p_1\rangle\}(dx)^4. \quad (9.93)$$

The next steps are similar to those of Section 9.2. The same partial integration is performed with $t_{\mu\nu}$, not producing an equal-time commutator; the equations are then decomposed into invariant kinematic forms and finally unsubtracted dispersion relations are used for the scalar parts to produce dispersion sum rules. We refer to the literature for the details: references (11), (76–77), (179–82), (244), (264). The results are partly equivalent to those of section 9.2, but new sum rules also appear.

† See in particular: (394).

These sum rules are probably less well justified than those of Section 9.2, for two reasons:

(a) The equal-time commutator also involves current space-components with each other, although they may be eliminated by appropriate antisymmetrization.

(b) The convergence properties of the dispersion relations are even more critical than in Section 9.2.

Independently of any detailed experimental information and any Regge-pole concepts Gourdin[181] was able to show that the assumption of unsubtracted dispersion relations in ν for all tensor amplitudes cannot be valid simultaneously for these sum rules and for those of Section 9.2. A certain cautiousness towards this assumption should be adopted in the light of the results of Section 9.5. At present it is not yet entirely clear how far unsubtracted dispersion relations should be accepted.

A second class is well represented by the sum rule of Drell and Hearn[118] on Compton scattering. The low-energy limit of Compton scattering has a semiclassical analogue in non-relativistic quantum mechanics, and one may express several low-energy values in terms of the charge and the magnetic moment of the target. Low[S 29] has investigated these relations more consistently and he has worked out the low-energy limits of Compton scattering amplitudes in terms of the target electromagnetic form factors.

Thus proposing energy dispersion relations in suitable amplitudes leads to sum rules for the low-energy values. In this way Drell and Hearn[118] derived their sum rule for the spin flip term in the forward Compton scattering amplitude, expressing the nucleon magnetic moments in terms of total cross-sections for the absorption of circularly polarized photons by protons polarized in parallel and antiparallel directions.

The connection of this approach to current algebras is not yet entirely clear; we refer to some related work: references (51–52), (118), (217), (221).

Finally the work of Amati, Jengo and Remiddi[25–26] deserves attention. So far, throughout Chapters 4 and 9, whenever dealing with the dispersion theory of current algebras, we have confined ourselves to dispersing in the energy variable ν. This seemed the safest choice by the scattering analogy of Section 4.1, although we have seen in Section 4.2 that other dispersion integrals involving variable values of q^2 are also useful in their connection to direct sum rules. The analogy of a current commutator integral to a scattering amplitude is a merely formal one. Scattering amplitudes for variable masses cannot easily be interpreted; but for current-commutator integrals this limitation does not exist. The work of Amati, Jengo and Remiddi has once again shown that fixed q^2 is not an indispensable limitation.

Note added in proof: The final part of chapter 9, and the short chapter 10, were included at proof stage. If this information had been available four months earlier, when the main text was written, the author would have economized in his presentation of the preceding sections.

The connection of their sum rules with Fubini's dispersion relations is best illustrated through early work of Bollini and Giambiagi[405,-6]. These authors apply the convolution theorem to the Fourier transform of the commutator of local currents

$$\delta(x_0)\left[j_0^\alpha\left(\frac{x}{2}\right), j_\mu^\beta\left(-\frac{x}{2}\right)\right] = if^{\alpha\beta\gamma}j_\mu^\gamma(0)\,\delta^4(x) \qquad (9.94)$$

and, defining

$$t_{0\mu}(Q_0,\vec{Q}) = \int e^{iQx}\left\langle p_2\left|\left[j_0^\alpha\left(\frac{x}{2}\right), j_\mu^\beta\left(-\frac{x}{2}\right)\right]\right|p_1\right\rangle (dx)^4 \qquad (9.95)$$

as in (9.26), they obtain the sum rule

$$\frac{1}{2\pi}\int t_{0\mu}(Q_0,\vec{Q})\,dQ_0 = if^{\alpha\beta\gamma}\,\langle p_2|\,j_\mu^\gamma(0)\,|p_1\rangle \qquad (9.96)$$

As it stands, (9.96) does not look manifestly covariant, but by suitable choices of the vectors $\vec{p_1}$, $\vec{p_2}$ and $\vec{Q}$ we can obtain special forms of (9.96) whose covariance is easily recognized.

The sum rules of section 9.2 are reproduced in the limit $(\vec{p_1}+\vec{p_2}) \to \infty$; $(\vec{p_1}-\vec{p_2})$ fixed; $\vec{Q}$ fixed. This is most easily seen by introducing intermediate states into (9.95) and following the arguments of sections 4.2 and 9.1. The interchange of the infinite-momentum limit with the Q_0-integration corresponds to the assumption of unsubtracted dispersion relations.

Instead of considering the matrix element of (9.94) between particle-states $\langle p_2|$ and $|p_1\rangle$, one may also consider it between the vacuum $\langle 0|$ and the particle–antiparticle pair state $|p_1,\bar{p}_2\rangle$. By crossing, one may equivalently assign a negative energy to $\langle p_2|$ and regard $(p_1+p_2)_\mu$ as space-like and $(p_2-p_1)_\mu$ as time-like. The

147

sum rules of Amati, Jengo and Remiddi[25,26] are derived by letting the spatial momentum of the state $|p_1,\bar{p}_2\rangle$, $(\overrightarrow{p_1+\bar{p}_2}) = (\overrightarrow{p_1-p_2})$, tend to infinity in (9.96). This limit selects the components proportional to Δ_μ and Δ_ν in (9.37); in analogy to the sum rules (9.47)-(9.51) we derive:

$$2if^{ijk}F^+(t) = \frac{-1}{2\pi}\int dv'b^{(2)}(v',s,t,w) \qquad (9.97)$$

$$2if^{ijk}F^+(t) = \frac{-1}{2\pi}\int dv'b^{(4)}(v',s,t,w) \qquad (9.98)$$

$$if^{ijk}F^-(t) = \frac{-1}{2\pi}\int dv'c^{(4)}(v',s,t,w) \qquad (9.99)$$

$$0 = \frac{1}{2\pi}\int dv'c^{(2)}(v',s,t,w) \qquad (9.100)$$

$$0 = \frac{1}{2\pi}\int dv'c^{(3)}(v',s,t,w) \qquad (9.101)$$

The integration variable v' is introduced as $v = \frac{1}{2}(q_2^2 - q_1^2)$, the fixed variable w as $w = \frac{1}{2}(q_1^2 + q_2^2)$. The appearance of these combinations can be understood either by going through the details of the derivation[26] or by noting that the crossing transformation $\langle p_2(p)| \Rightarrow |\bar{p}_2(-p)\rangle$ changes the dispersion variable $v = \frac{1}{4}(p_1+p_2)(q_1+q_2)$ into $\frac{1}{4}(p_1-p_2)(q_1+q_2) = \frac{1}{4}(q_2-q_1)(q_1+q_2) = 2v$. This remark is not to be misunderstood as suggesting that the sum rules (9.97)–(9.101) should be regarded as analytic continuations of (9.47)–(9.51), in fact the integrands receive contributions from quite a different class of intermediate states; they consist of terms

$$\sum_n \langle 0|j_\mu^j(0)|n\rangle\langle n|j_\nu^j|a(p_1),\bar{b}(-p_2)\rangle$$

$$\text{and} \qquad (9.102)$$

$$\sum_n \langle 0|j_\nu^j(0)|n\rangle\langle n|j_\mu^i|a(p_1),\bar{b}(-p_2)\rangle$$

Only states of spin 0 or 1 and definite parity can contribute.

As in section 9.4 we may again use the independence of the commutator on the variables s and w to derive superconvergence relations, this time for form factors of currents as a function of momentum transfer. The full consequences of the theory are not yet worked out, but an interesting result has already been achieved, namely a strong damping of the nucleon magnetic moment form factor $F_2^{\text{el}}(\Delta^2)$ (see equation (1.3)) at high momentum transfer

$$\lim_{\Delta^2 \to \infty} [\Delta^2\, F_2^{\text{el}}(\Delta^2)] = 0. \tag{9.103}$$

This asymptotic result seems to be correct experimentally, but theoretically it is so far unexplained.

At this place we wish to mention studies of further classes of sum rules derived from (9.95). References: (443), (444), (502), (503), (511).

Both in deriving the sum rules (9.47)–(9.51) and (9.97)–(9.101) we have always interchanged the infinite-momentum limit with the summation over intermediate states, or equivalently we have assumed unsubtracted dispersion relations. As already discussed in section 3.2, these steps imply the assumption that disconnected configurations as illustrated on page 40 can be neglected, because in the infinite-momentum limit these states would contribute at an infinite value of their invariant mass and have no natural place in the sum rules. Recently these contributions have been investigated carefully by Amati, Jengo and Remiddi[398]. We state only two of their main results here:

1. The importance of the different classes of contributions depends very much on the details of the sum rule. All kinds are important in (9.95). Terms like

$$\sum_{n'} \langle b(p_2)|j_\mu^i(0)|b(p_2); n' \rangle \langle n'; b(p_2)|j_\nu^j(0)|a(p_1) \rangle$$
$$= \sum_{n} \langle 0|j_\mu^i(0)|n' \rangle \langle n'; b(p_2)|j_\nu^j(0)|a(p_1) \rangle \tag{9.104}$$

which have been disregarded in the limit $(p_1 + p_2) \to \infty$ (section 3.2) provide the "natural contributions" in the limit $(p_1 - p_2) \to \infty$, see (9.102). On the other hand the connected terms kept in section 3.2 are pushed to infinite masses in the latter limit.

2. Evidently one should become more careful about disregarding disconnected terms or subtractions, as has already been emphasized throughout section 9.5. A subtraction constant in the dispersion variable may of course be a function of the remaining variables. The superconvergence relations in section 9.4, for instance, would no longer be valid with subtraction functions $s(t, q_1^2, q_2^2)$ added in eqns (9.47)–(9.53)

$$F(t) = \frac{1}{2\pi} \int dv' f(v',t,q_1^2,q_2^2) + s(t,q_1^2,q_2^2). \qquad (9.105)$$

Proposing to identify $s(t,q_1^2,q_2^2)$ with the contributions from disconnected intermediate states, Amati, Jengo and Remiddi[398] could explicitly demonstrate a cancellation of the singularities in q_1^2 and q_2^2 on the right hand side of (9.105). This means that superconvergence (see (9.73)) is not a simple consequence of locality of the commutator, but also a dynamical statement on the subtraction term.

The balance of the t-singularities in (9.105) is not yet completely understood. A connection of the singularities of $F(t)$ with the divergence of the integral due to Regge behaviour of the integrand, as conjectured by Fubini and Segrè[139] is under investigation in a more refined Regge analysis, taking account of the helicity structure of the tensor amplitudes: (414), (415), (498). With the present knowledge, a non-vanishing current commutator seems to require a fixed pole at $J = 1$ in the complex angular momentum plane of $T_{\mu\nu}$ (9.22). In this respect, $T_{\mu\nu}$ is different from a hadron scattering amplitude; it can be described either as a second order lepton-hadron scattering amplitude or an off mass shell hadron amplitude. A discussion of the details would require a substantially higher standard of Regge pole theory than the rest of this book, so we refer the interested reader to the literature: (394), (397), (398), (414), (415), (492), (498), (502).

CHAPTER 10

Possible Connections of Current Algebras and Dynamics

In the last three chapters we have reached the fields of present research in current algebras. Assumptions have been put forward, examined, criticized, partly invalidated or confirmed. In our review we tried to discuss the present problems without elaborating on possibly short-lived hypothesis. Further developments may well correct our approach.

With these concluding remarks we try to hint at some fields for future research and to give some guidance to the partial answers given so far in the literature.

Current algebras may be introduced as a postulate of commutation relations, generalizing the quantum mechanical commutators of Heisenberg and Dirac to observables which are no longer purely kinematical; but it may also be more fruitful to elaborate on gauge transformations, as indicated by Nauenberg[277] and continued in the interpretation of Veltman's equations (see section 8.2). It may be correct to regard current algebras as based on strong interactions and to connect them to bootstrap theories, but it may also be impossible to understand current algebras without considering all interactions on the same level.

Numerous and diverse attempts have been made to go beyond the postulate of commutators and to derive current algebras or their results from other theoretical concepts: (94), (109), (110), (246), (247), (364), (365), (380), (434), (445), (501), (508). Among different possible lines we mention in particular the work of Dashen and Frautschi[109],[110], of Taha[501] and of Dietz and Kupsch[428] who set out to prove that the commutator of a charge and a vector or axial vector current produces again a vector or axial vector current. To arrive at such a statement from more general principles one should no longer accept the currents

as given by observation, but define currents by specified properties and require charge-current commutators to have the same
properties.

In the pioneer papers of Dashen and Frautschi these concepts
are not yet clearly separated, among the followers Dietz and
Kupsch[428] demonstrated that a charge-current commutator
has the correct Lorentz-transformation properties as a vector
and is free from covariant Schwinger terms under sufficiently
general conditions, and Taha[501] showed that it has the singularity structure suggested by linearized unitarity relations.
In these arguments assumptions are made mainly about regularity
of distributions in x-space or, almost equivalently, suitable
asymptotic behaviour in p-space.

At the present stage of the theory we feel that current algebras
provide some kind of general framework which may contain
enough freedom to accommodate quite different dynamical
situations.[397] Respective implications of current algebras
and dynamical theories on each other may be a promising field
for further research.

There is sufficient evidence now that current algebras may be
true; next we would like to understand why they should be true.

Some Details on LSZ Reduction Formulae

LAGRANGIAN field theories of strong interactions have not been developed to the stage where specific predictions can be compared with experiments. This is why we cannot say anything about the validity of particular field-theoretic equations for strong interactions and also why the general scheme of Lagrangian field theories has lost some interest over the last ten years. Nevertheless there are some model-independent principles and methods derived from field theory which are of more general use.

One of these principles is the representation of scattering amplitudes by LSZ (Lehmann–Symanzik–Zimmermann) reduction formulae. We will only derive a very special type which we will need in particular. We will work in the Heisenberg picture where the time-dependent field operators associated with the particles satisfy an asymptotic condition; that is they behave like free field operators in the limits $t \to \pm\infty$, at least for suitably dispersed states where the particles are asymptotically out of each other's range of force. For illustration we consider pions. We have pion field operators: $\varphi(x)$, which for free particles $[\varphi^f(x)]$ may be expanded into creation and annihilation operators for states characterized by wave functions $f_\alpha(x)$:

$$\varphi^f(x) = \sum_\alpha (f_\alpha(x)\varphi^\alpha + f_\alpha^+(x)\varphi^{\alpha+}). \tag{A1}$$

The wave functions are solutions of the Klein–Gordon equation

$$(\mu^2 + \Box)f_\alpha(x) = 0 \tag{A2}$$

which $\varphi^f(x)$ obeys as well. The orthogonality relation for the wave functions follows from the properties of the Klein–Gordon equation:

153

$$i \int (dx)^3 \left[f_\beta^+(x) \frac{\partial f_\alpha(x)}{\partial x_0} - \frac{\partial f_\beta^+(x)}{\partial x_0} f_\alpha(x) \right]$$

$$\stackrel{\text{def}}{=} i \int (dx)^3 \left(f_\beta^+ \overset{\leftrightarrow}{\frac{\partial}{\partial x_0}} f_\alpha \right) = \delta_{\alpha\beta}, \qquad \text{(A3)}$$

$f_\alpha(x)$ contains only negative frequences ($e^{-ik_0 x_0}$), $f_\alpha^+(x)$ only positive ones, because only states of positive energy are admitted. Consequently we construct projection operators

$$\varphi^\alpha = -i \int (dx)^3 \left[\varphi^f(x) \overset{\leftrightarrow}{\frac{\partial}{\partial x_0}} f_\alpha^+(x) \right]; \quad \langle \alpha | = \langle 0 | \varphi^\alpha, \qquad \text{(A4)}$$

$$\varphi^{\alpha+} = i \int (dx)^3 \left[\varphi^f(x) \overset{\leftrightarrow}{\frac{\partial}{\partial x_0}} f_\alpha(x) \right]; \quad | \alpha \rangle = \varphi^{\alpha+} | 0 \rangle. \qquad \text{(A5)}$$

$f_\alpha(x)$ are localized states, but at the end of the calculations it is convenient to take plane waves of momentum k^α: $f_\alpha(x) \rightarrow e^{-ik_\alpha x}$ with positive energy $k_0^\alpha > 0$.

For interacting fields we have $(\mu^2 + \Box)\varphi(x) \neq 0$ and the expansion (A1) is no longer generally valid. In assuming the asymptotic condition we suppose that for $t \rightarrow \pm \infty$ with suitably dispersed states the projections (A4) and (A5) with the free field $\varphi^f(x)$ replaced by $\varphi(x)$ again produce creation and annihilation operators for asymptotically free states.

An S-matrix element is represented with time-independent Heisenberg states: $_{+\infty}\langle p_2 q_2 | p_1 q_1 \rangle_{-\infty}$, where the subscripts indicate the asymptotic limits. At $t \rightarrow + \infty$ $\langle p_2 q_2 |$ is found to consist of a pion with momentum q_2 and some other particle with momentum p_2. We can therefore represent

$$\langle p_2 q_2 | p_1 q_1 \rangle = \lim_{x_0 \rightarrow +\infty} -i \int (dx)^3 [\langle p_2 | \varphi(x) | p_1 q_1 \rangle \overset{\leftrightarrow}{\frac{\partial}{\partial x_0}} f_{q_2}^+(x)] \qquad \text{(A6)}$$

using the projection (A4). We rewrite (A6)

$$\langle p_2 q_2 | p_1 q_1 \rangle = -i \int (dx)^4 \frac{\partial}{\partial x_0} \left[\langle p_2 | \varphi(x) | p_1 q_1 \rangle \overset{\leftrightarrow}{\frac{\partial}{\partial x_0}} f_{q_2}^+(x) \right]$$

$$-i \lim_{x_0 \rightarrow -\infty} \int (dx)^3 \left[\langle p_2 | \varphi(x) | p_1 q_1 \rangle \overset{\leftrightarrow}{\frac{\partial}{\partial x_0}} f_{q_2}^+(x) \right] \qquad \text{(A7)}$$

The second term in (A7) gives a contribution only if $| q_1 \rangle$ is the same state as $\langle q_2 |$; then the projection of $\varphi(x)$ acts as annihilation operator, and we obtain the disconnected unit part of the S-matrix: $\delta_{q_1 q_2} \langle p_2 | p_1 \rangle$. This part is uninteresting for us, so we continue to investigate the first term and exclude the second one by assuming $\langle p_2|p_1\rangle = 0$.

$$\langle p_2 q_2|p_1 q_1\rangle = -i \int (dx)^4 \left[\langle p_2|\varphi(x)|p_1 q_1\rangle \frac{\partial^2}{\partial x_0^2} f_{q_2}^+(x) \right.$$
$$\left. -f_{q_2}^+(x)\frac{\partial^2}{\partial x_0^2}\langle p_2|\varphi(x)|p_1 q_1\rangle \right]. \tag{A8}$$

We replace $\left[\frac{\partial^2}{\partial x_0^2} f_{q_2}^+(x)\right]$ by (A2) and use the theorem of Gauss for space derivatives:

$$\langle p_2 q_2|p_1 q_1\rangle = i \int (dx)^4 f_{q_2}^+(x)(\Box + \mu^2)\langle p_2|\varphi(x)|p_1 q_1\rangle. \tag{A9}$$

A more interesting formula is obtained if the reduction technique is applied a second time on $\langle p_2|g(x)|p_1 q_1\rangle$ where $g(x)$ is any local operator, to be identified with $\varphi(x)$ if required.

$$\langle p_2|g(x)|p_1 q_1\rangle = \lim_{y_0\to-\infty} i \int (dy)^3 \left[\langle p_2|g(x)\varphi(y)|p_1\rangle \overleftrightarrow{\frac{\partial}{\partial y_0}} f_{q_1}(y) \right]$$
$$= \lim_{y_0\to-\infty} i \int_{y_0} (dy)^3 \left[\vartheta(x_0-y_0)\langle p_2|[g(x),\varphi(y)]|p_1\rangle \overleftrightarrow{\frac{\partial}{\partial y_0}} f_{q_1}(y) \right]. \tag{A10}$$

The introduction of the commutator $[g(x),\varphi(y)]$ into (A10) instead of the operator product does not change anything because

$$\lim_{y_0\to-\infty} i \int (dy)^3 \langle p_2|\varphi(y)g(x)|p_1\rangle \overleftrightarrow{\frac{\partial}{\partial y_0}} f_{q_1}(y) = 0. \tag{A11}$$

In the projection of (A11) the field operator $\varphi(y)$ acts as annihilation operator on the state $\langle p_2|$, which, though constructed to represent an asymptotically free particle at $t=+\infty$, will also represent it at $t=-\infty$, because a single-particle state is not affected by interactions in a renormalized field theory.

More critical is the factor $\vartheta(x_0-y_0)$; its inclusion or exclusion from the time derivative $\partial/\partial y_0$ leads to an ambiguity which will be displayed below. Now we apply a similar trick as before:

$$\langle p_2|g(x)|p_1 q_1\rangle$$
$$= -i \int (dy)^4 \frac{\partial}{\partial y_0}\left[\vartheta(x_0-y_0)\langle p_2|[g(x),\varphi(y)]|p_1\rangle \overleftrightarrow{\frac{\partial}{\partial y_0}} f_{q_1}(y) \right] \tag{A12}$$

where the factor $\vartheta(x_0-y_0)$ disposes of the term at $y_0=+\infty$. Similarly as with (A8) we derive:

$$\langle p_2 \mid g(x) \mid p_1 q_1 \rangle$$

$$= i \int f_{q_1}(y)\,(\Box_y + \mu^2)\,[\vartheta(x_0 - y_0)\,\langle p_2 \mid [g(x), \varphi(y)] \mid p_1\rangle]\,(dy)^4$$

$$\left[\left[-i \int f_{q_1}(y)\,(dy)^4 \frac{\partial}{\partial y_0}\{\delta(x_0 - y_0)\,\langle p_2 \mid [g(x), \varphi(y)] \mid p_1\rangle\}\right]\right]. \quad (A13)$$

$\overset{\leftrightarrow}{}$ The second term in (A13) disappears, if in (A10) the operation $\partial/\partial y_0$ is supposed to include $\vartheta(x_0 - y_0)$; otherwise it appears. We regard this ambiguity as a consequence of the poor mathematical definition of products of distributions and we will drop the term as there is no evidence in its favour. It might be connected to subtraction constants. We summarize

$$\langle p_2 q_2 \mid p_1 q_1 \rangle = - \int f_{q_2}^+(x) f_{q_1}(y)\,(\Box_x + \mu^2)(\Box_y + \mu^2)\,[\vartheta(x_0 - y_0)$$

$$\langle p_2 \mid [\varphi(x), \varphi(y)] \mid p_1\rangle]\,(dx)^4 (dy)^4. \quad (A14)$$

Now we may insert $f_{q_1}(y) = e^{-iq_1 y}$, $f_{q_2}^+(x) = e^{iq_2 x}$ and, with suitable imaginary parts attached to q_{10} and q_{20}, we may rewrite (A14) further, using a translation by y and defining $z = x - y$

$$\langle p_2 q_2 \mid p_1 q_1 \rangle = i (2\pi)^4 \delta^4(p_1 + q_1 - p_2 - q_2)\,T(p_1, q_1, p_2, q_2) \quad (A15)$$

$$T(p_1, q_1, p_2, q_2) = i \int e^{iq_2 z}(\mu^2 - q_2^2)(\mu^2 - q_1^2)\vartheta(z_0)$$

$$\langle p_2 \mid [\varphi(z), \varphi(0)] \mid p_1\rangle\,(dz)^4. \quad (A16)$$

To have non-vanishing matrix elements in (A16) the integral

$$\int e^{iq_2 z}\vartheta(z_0)\,\langle p_2 \mid [\varphi(z), \varphi(0)] \mid p_1\rangle\,(dz)^4$$

must have poles at $q_1^2 = \mu^2$, $q_2^2 = \mu^2$; in (A16) their residue is projected.

S-matrix elements are defined only on the mass shell of the scattered particles; but the integral may be given a meaning also at $q_1^2 \neq \mu^2$, $q_2^2 \neq \mu^2$, provided there is a way to specify the interpolating field $\varphi(x)$. This interpolation is far from unique. Any local operator with the quantum numbers of the particle in question can serve as an interpolating field, if it has the pole at the particle mass required in (A16) with a correctly factorizing residue. In the case of a pion this is expected to be true with the divergence of the axial vector current. For details we refer to Section 6.1.

Conventions

1. In this book we always use *time-like metric*: $g_{00} = 1$, $g_{ii} = -1$ ($i = 1,2,3$). The scalar product is always expressed as $a \cdot b = a_0 b_0 - (\bar{a} \cdot \bar{b})$; vector space-components are denoted by a bar, wherever an ambiguity may arise.

2. Our *units* are chosen such that $\hbar = c = 1$, so these constants will always be omitted. So the Schrödinger equation takes the form $i(d/dt)|\psi\rangle = H|\psi\rangle$ and consequently we have $\langle f|\hat{O}(x)|i\rangle = e^{i(p^f - p^i)x}\langle f|\hat{O}(0)|i\rangle$ for matrix elements of local operators.

All equations have the dimension of a positive or negative power of a mass under these conditions; length or time has the dimension m^{-1}.

3. States will be *normalized* invariantly: $\langle f|i\rangle = (2\pi)^3 2E_p \delta^3(p^f - p^i)$; if occasionally unit normalization is more convenient we denote unit-normalized states by double brackets $\langle\!\langle f|i\rangle\!\rangle = \delta^3(p^f - p^i)$.

4. *γ-matrices* satisfying $(\gamma_\mu \gamma_\nu + \gamma_\nu \gamma_\mu) = 2g_{\mu\nu}$ will be chosen such that γ_0 is hermitean, γ_i are antihermitean: $(\gamma_i)^+ = -\gamma_i$. If we need a special representation we will take

$$\gamma_0 = \begin{pmatrix} 1 & \\ & -1 \end{pmatrix}; \quad \gamma_i = \begin{pmatrix} & -\sigma_i \\ \sigma_i & \end{pmatrix} \tag{C1}$$

with the Pauli matrices σ_i.

We introduce the σ-matrices as $\sigma_{\mu\nu} = \tfrac{1}{2}(\gamma_\mu \gamma_\nu - \gamma_\nu \gamma_\mu)$ and $\gamma^5 = -i\gamma_0 \gamma_1 \gamma_2 \gamma_3$ as a hermitean matrix; in the special representation (C1) we have $\gamma^5 = \begin{pmatrix} & 1 \\ 1 & \end{pmatrix}$.

5. The *Dirac equation* then takes the form

$$\left(+i\gamma_\mu \frac{\partial}{\partial x_\mu} - m\right)u(p) = 0 \qquad \text{equivalently } (\gamma \cdot p - m)u(p) = 0.$$

157

The spinors are normalized $u^+(p)\gamma_0 u(p) = 2m$; for $u^+(p)\gamma_0$ we will frequently use the notation $\bar{u}(p)$.

A neutrino spinor [of negative helicity] satisfies $(1+\gamma^5)u_\nu(p) = 2u_\nu(p)$.

6. This normalization leads to the following *polarization sums*:

$$\sum_{(P)} u(p)\bar{u}(p) = \gamma p + m \qquad \text{for spinor states,}$$

$$\sum_{(P)} \epsilon_\mu(p)\epsilon_\nu(p) = -g_{\mu\nu} + \frac{p_\mu p_\nu}{m^2} \qquad \text{for vector states.}$$

Bibliography on Current Algebras

The literature on current algebras contains now (April 1967) about 500 contributions; there will probably be more than 600 when this book is published.

I tried to collect a large number of references, but I do not claim that my list is exhaustive, in particular not as far as preprints are concerned.

I apologize to all authors who find some of their contributions not listed because I did not become aware of them among the almost exploding literature on current algebras.

Contributions of mainly review character are denoted by (R).

1. H. D. I. ABABARNEL, C. G. CALLAN and S. B. TREIMAN: Princeton University preprint: *Connection between Strong and Weak Interactions from Algebra of Commutators*.
2. M. ADEMOLLO: *Nuovo Cimento* **46A**, 156 (1966).
3. M. ADEMOLLO and R. GATTO: *Nuovo Cimento* **44A**, 282 (1966).
4. M. ADEMOLLO, R. GATTO, G. LONGHI and G. VENEZIANO: *Phys. Lett.* **22**, 521 (1966).
5. M. ADEMOLLO, R. GATTO, G. LONGHI and G. VENEZIANO: *Phys. Rev.* **153**, 1623 (1967).
6. S. L. ADLER: *Phys. Rev.* **135B**, 963 (1964).
7. S. L. ADLER: *Phys. Rev.* **137B**, 1022 (1965).
8. S. L. ADLER: *Phys. Rev.* **139B**, 1638 (1965).
9. S. L. ADLER: *Phys. Rev. Lett.* **14**, 1051 (1965).
10. S. L. ADLER: *Phys. Rev.* **140B**, 736 (1965).
11. S. L. ADLER: *Phys. Rev.* **143**, 1144 (1966). E **151**, 1342 (1966).
12. (R) S. L. ADLER: *Proceedings of the Int. Conf. on Weak Interactions, Argonne 1965*, p. 285.
13. S. L. ADLER and C. G. CALLAN: CERN preprint.
14. S. L. ADLER and F. J. GILMAN: *Phys. Rev.* **152**, 1460 (1966).
15. S. L. ADLER and F. J. GILMAN: Caltech preprint: CALT-68–99.
16. M. A. AHMED: Durham preprint: *The Commutation of Axial Charge with Axial Divergence*.
17. C. H. ALBRIGHT: *Phys. Lett.* **24B**, 100 (1967).
18. V. A. ALESSANDRINI, M. A. BÉG and L. S. BROWN: *Phys. Rev.* **144**, 1137 (1966).
19. V. DE ALFARO, S. FUBINI, G. FURLAN and C. ROSSETTI: *Phys. Lett.* **21**, 576 (1966).
20. W. ALLES and R. JENGO: *Nuovo Cimento* **42A**, 419 (1966).
21. G. ALTARELLI: *Nuovo Cimento* **45A**, 426 (1966).

Bibliography

22. G. ALTARELLI, R. GATTO, L. MAIANI and G. PREPARATA: *Phys. Rev. Lett.* **16**, 918 (1966).
23. D. AMATI and S. BERGIA: *Nuovo Cimento* **45A**, 15 (1966).
24. D. AMATI, C. BOUCHIAT and J. NUYTS: *Phys. Lett.* **19**, 59 (1965).
25. D. AMATI, R. JENGO and E. REMIDDI: *Phys. Lett.* **22**, 674 (1966).
26. D. AMATI, R. JENGO and E. REMIDDI: CERN preprint TH 720.
27. V. AMBEGAOKAR and P. TARJANNE: *Phys. Lett.* **20**, 535 (1966).
28. M. ANIS ALAM: *Nuovo Cimento* **46A**, 762 (1966).
29. J. ARAFUNE, Y. JWASAKI, K. KIKKAWA, S. MATSUDA and K. NAKAMURA: *Phys. Rev.* **143**, 1220 (1966).
30. R. ARNOWITT: *Nuovo Cimento* **40A**, 985 (1965).
31. P. R. AUVIL and N. G. DESHPANDE: *Nuovo Cimento* **46A**, 766 (1966).
32. I. G. AZNAURYAN and L. D. SOLOVIEV: Dubna preprint E-2544.
33. P. BABU: *Phys. Rev.* **148**, 1440 (1966).
34. S. BADIER and C. BOUCHIAT: *Phys. Lett.* **20**, 529 (1966).
35. P. DE BAENST, M. KONUMA and J. WEYERS: *Nuovo Cimento* **45A**, 501 (1966).
36. P. DE BAENST, D. SPEISER and J. WEYERS: *Phys. Lett.* **21**, 100 (1966).
37. D. BAILIN: Sussex preprint: *S and P currents and the Nonleptonic Hyperon Decays*.
38. M. BAKER: *JETP Lett.* **4**, 157 (1966).
39. A. P. BALACHANDRAN, M. G. GUNDZIG and F. NICODEMI: *Nuovo Cimento* **44A**, 1257 (1966).
40. A. P. BALACHANDRAN, M. G. GUNDZIG and F. NICODEMI: *Proc. of the Boulder Conference on Particle Physics (1966)* — lecture by A. P. BALACHANDRAN.
41. A. P. BALACHANDRAN, M. G. GUNDZIG and F. NICODEMI: Inst. Adv. St. preprint: *Current Commutators and Mass Extrapolations; Theory and Applications* and Addendum.
42. A. P. BALACHANDRAN, M. G. GUNDZIG and S. PAKVASA: *Phys. Rev.* **153**, 1553 (1967).
43. A. P. BALACHANDRAN and F. NICODEMI: Syracuse preprint NYO-3399–53.
44. A. P. BALACHANDRAN and H. PIETSCHMANN: *Nucl. Phys.* **43**, 321 (1963).
45. K. BARDAKCI: Berkeley preprint: *Sidewise Dispersion Relations and the Axial Vector Coupling Constant*.
46. K. BARDAKCI, M. B. HALPERN and G. SEGRÈ: Berkeley preprint: *Ward Identities, Unsubtracted Dispersion Relations and Feynman Graphs*.
47. I. M. BAR-NIR: *Phys. Rev. Lett.* **16**, 473 (1966); E **16**, 927.
48. K. J. BARNES and E. KAZES: *Phys. Rev. Lett.* **17**, 978 (1966).
49. A. BARTL and C. P. KORTHALS ALTES: *Nuovo Cimento* **47A**, 721 (1967).
50. A. BECKERS: *Nuovo Cimento* **44A**, 822 (1966).
51. M. A. B. BÉG: *Phys. Rev. Lett.* **17**, 333 (1966).
52. M. A. B. BÉG: *Phys. Rev.* **150**, 1276 (1966).
53. (R) J. S. BELL: *Proc. of the 1966 CERN School of Physics*, CERN Report 66–29.
54. J. S. BELL: *Nuovo Cimento* **47A**, 616 (1967).
55. J. S. BELL: CERN preprint TH 725.
56. J. S. BELL and S. N. BERMAN: *Nuovo Cimento* **47A**, 807 (1967).
57. S. BERGIA and F. G. LANNOY: CERN preprint.
58. A. BIETTI: *Phys. Rev.* **140B**, 908 (1965).
59. A. BIETTI: *Phys. Rev.* **142**, 1258 (1966).
60. A. BIETTI: *Phys. Rev.* **144**, 1289 (1966).

61. A. Bietti: *Nuovo Cimento* **44A**, 553 (1966).
62. A. M. Bincer: *Phys. Rev. Lett.* **16**, 754 (1966).
63. H. Biritz and H. Pietschmann: *Acta Physica Austriaca* **18**, 252 (1964).
64. S. N. Biswas, A. Kumar and P. Saxena: *Phys. Rev. Lett.* **17**, 268 (1966).
65. S. N. Biswas and S. H. Patil: University of California preprint: *Meson Decays in Broken SU_3 and Current Commutation Relations.*
66. J. D. Björken: *Phys. Rev. Lett.* **16**, 408 (1966).
67. J. D. Björken: *Phys. Rev.* **148**, 1467 (1966).
68. (R) S. A. Bludman: Lectures at École d'été de physique théorique — Cargèse, 1966.
69. N. N. Bogolubov, V. A. Matvejev and A. N. Tavkhelidze: *Nuovo Cimento* **48A**, 132 (1967).
70. M. Boiti and C. Rebbi: *Nuovo Cimento* **43A**, 214 (1966).
71. S. Bose and S. Bose: *Phys. Rev.* **148**, 1518 (1966).
72. S. K. Bose and S. N. Biswas: *Nuovo Cimento* **43A**, 1182 (1966).
73. S. K. Bose and S. N. Biswas: *Phys. Rev. Lett.* **16**, 330 (1966).
74. S. K. Bose and Y. Hara: *Phys. Rev. Lett.* **17**, 409 (1966).
75. S. K. Bose and A. H. Zimerman: *Nuovo Cimento* **43A**, 1165 (1966). E**45A**, 1056 (1966).
76. C. Bouchiat and Ph. Meyer: *Nuovo Cimento* **44A**, 843 (1966).
77. C. Bouchiat and Ph. Meyer: *Nuovo Cimento* **45A**, 108 (1966).
78. C. Bouchiat and Ph. Meyer: *Phys. Lett.* **22**, 198 (1966).
79. L. S. Brown: *Phys. Rev.* **150**, 1338 (1966).
80. L. S. Brown and C. M. Sommerfield: *Phys. Rev. Lett.* **16**, 751 (1966).
81. F. Buccella, M. De Maria, M. Lusignoli, B. Taglienti and G. Violini: *Nuovo Cimento* **45A**, 1054 (1966).
82. F. Buccella, G. Veneziano and R. Gatto and S. Okubo: *Phys. Rev.* **149**, 1268 (1966).
84. (R) N. Cabibbo: Lecture at the Third Coral Gables Conference on Symmetry Principles at High Energy, 1966.
85. (R) N. Cabibbo: Lecture at the XIII Internat. Conf. on High Energy Physics, Berkeley, 1966.
86. N. Cabibbo, L. Horwitz and Y. Ne'eman: *Phys. Lett.* **22**, 336 (1966).
87. N. Cabibbo, L. Horwitz, J. J. J. Kokkedee and Y. Ne'eman: *Nuovo Cimento* **45A**, 275 (1966).
88. N. Cabibbo and L. A. Radicati: *Phys. Lett.* **19**, 697 (1966).
89. N. Cabibbo and H. Ruegg: *Phys. Lett.* **22**, 85 (1966).
90. C. G. Callan and S. B. Treiman: *Phys. Rev. Lett.* **16**, 153 (1966).
91. G. Calucci and G. Denardo: Trieste University preprint: *An Estimate of Second Class Form Factors in Semileptonic Decays of Hyperons.*
92. G. Calucci, G. Denardo and C. Rebbi: Trieste University preprint: *Symmetry Breaking Corrections to Semileptonic Decays of Baryons.*
93. E. Celeghini and R. Gatto: *Nuovo Cimento* **43A**, 219 (1966).
94. Chan Hong-Mo: *Nuovo Cimento* **43A**, 347 (1966).
95. W. K. Cheng and C. W. Kim: Pennsylvania preprint: *Saturation of Sum Rules in Particle Representation of Resonances.*
96. Chia-Hwa Chan: *Phys. Lett.* **20**, 70 (1966).
97. Y. T. Chiu and J. Schechter: *Phys. Rev. Lett.* **16**, 1022 (1966).
98. Y. T. Chiu, J. Schechter and Y. Ueda: *Phys. Rev.* **150**, 1201 (1966).
99. Y. T. Chiu, J. Schechter and Y. Ueda: Chicago preprint: EFINS 66–74.
100. L. J. Clavelli: Chicago preprint E FINS 66–73.
101. E. R. McCliment: Max Planck Institut München preprint: *Current*

Algebra and Broken Symmetric Relations between Axial Vector Coupling Constants.

102. G. COCHO and HARUN AR RASHID: *Prog. Theor. Phys.* **36**, 1150 (1966).
103. F. COESTER: Argonne preprint: *Current Algebras and Configuration Mixing.*
104. F. COESTER and G. ROEPSTORFF: Argonne preprint: *Current Algebras at Infinite Momentum.*
105. S. COLEMAN: *Journal of Math. Phys.* **7**, 787 (1966).
106. S. COLEMAN: *Phys. Lett.* **19**, 144 (1966).
107. J. M. CORNWALL: *Phys. Rev. Lett.* **16**, 1174 (1966).
108. G. COSTA and A. H. ZIMERMAN: *Nuovo Cimento* **46A**, 198 (1966).
109. R. F. DASHEN and S. C. FRAUTSCHI: *Phys. Rev.* **143**, 1171 (1966).
110. R. F. DASHEN and S. C. FRAUTSCHI: *Phys. Rev.* **145**, 1287 (1966).
112. R. F. DASHEN and M. GELL-MANN: *Phys. Lett.* **17**, 145 (1965).
113. R. F. DASHEN and M. GELL-MANN: *Phys. Lett.* **17**, 148 (1965).
114. R. F. DASHEN and M. GELL-MANN: Lecture at the Third Coral Gables Conference on Symmetry Principles at High Energies, 1966.
115. R. F. DASHEN and M. GELL-MANN: *Phys. Rev. Lett.* **17**, 340 (1966).
116. K. DIETZ and W. DRECHSLER: *Nuovo Cimento* **40**, 634 (1965).
117. E. DONINI, F. PEMPINELLI and S. SCIUTO: *Nuovo Cimento* **46A**, 142 (1966).
118. S. D. DRELL and A. C. HEARN: *Phys. Rev. Lett.* **16**, 908 (1966).
119. T. EBATA: Ohio preprint: COO-1545–6.
120. D. K. ELIAS and J. C. TAYLOR: *Nuovo Cimento* **44A**, 518 (1966).
121. B. d'ESPAGNAT and J. ILIOPOULOS: *Phys. Lett.* **21**, 232 (1966); E **21**, 719.
122. E. FABRI and L. E. PICASSO: *Phys. Rev. Lett.* **16**, 408 (1966).
123. E. FABRI, L. E. PICASSO and F. STROCCHI: Pisa preprint: *Broken Symmetries in Quantum Field Theory.*
124. R. FAUSTOV: *Nuovo Cimento* **45A**, 145 (1966).
125. R. N. FAUSTOV, R. E. KALLOSH and V. G. PISARENKO: Dubna preprint E 2865.
126. FAYYAZUDDIN and RIAZUDDIN: *Nuovo Cimento* **47A**, 222, (1967).
127. FAYYAZUDDIN, RIAZUDDIN and M. S. K. RAZMI: *Phys. Rev.* **141**, 1509 (1966).
128. E. FERRARI, V. S. MATHUR and L. K. PANDIT: *Phys. Lett.* **21**, 560 (1966).
129. R. P. FEYNMAN, M. GELL-MANN and G. ZWEIG, *Phys. Rev. Lett.* **13**, 678 (1964).
130. P. H. FRAMPTON: Oxford preprint: *The Chirality Commutator and Vector Mesons.*
131. P. H. FRAMPTON and J. C. TAYLOR: Oxford preprint: *Superconvergence Sum Rules in $\rho\pi$ Scattering.*
132. S. FUBINI: *Nuovo Cimento* **43A**, 475 (1966).
133. (R) S. FUBINI: Lecture at the New York Meeting of the American Physical Society, 1966.
134. (R) S. FUBINI: Lectures at the École d'été de physique théorique, Cargèse, 1966, and at the International Advanced Study Institute, Istanbul, 1966.
135. S. FUBINI and G. FURLAN: *Physics* **1**, 229 (1965).
136. S. FUBINI, G. FURLAN and C. ROSSETTI: *Nuovo Cimento* **40A**, 1171 (1965).
137. S. FUBINI, G. FURLAN and C. ROSSETTI: ICTP Trieste preprint IC/65/58.
138. S. FUBINI, G. FURLAN and C. ROSSETTI: *Nuovo Cimento* **43A**, 161 (1966).
139. S. FUBINI and G. SEGRÈ: *Nuovo Cimento* **45A**, 641 (1966).
140. S. FUBINI and G. SEGRÈ and J. D. WALECKA: *Ann. Phys.* **39**, 381 (1966).
141. N. H. FUCHS: *Phys. Rev.* **149**, 1145 (1966).

142. N. H. FUCHS: *Phys. Rev.* **150**, 1241 (1966).
143. N. H. FUCHS: Purdue preprint: Applications of Current Commutator Sum Rules.
144. Y. FUJII: *Phys Rev. Lett.* **17**, 613 (1966).
145. Y. FUJII: Stanford preprint ITP 211.
146. G. FURLAN, R. JENGO and E. REMIDDI: *Phys. Lett.* **20**, 679; E **21**, 720 (1966).
147. G. FURLAN, R. JENGO and E. REMIDDI: *Nuovo Cimento* **44A**, 427 (1966).
148. G. FURLAN, F. G. LANNOY, C. ROSSETTI and G. SEGRÈ: *Nuovo Cimento* **38**, 1747 (1965).
149. G. FURLAN, F. G. LANNOY, C. ROSSETTI and G. SEGRÈ: *Nuovo Cimento* **40A**, 597 (1965).
150. G. FURLAN and B. RENNER: *Nuovo Cimento* **44A**, 536 (1966).
151. R. G. FURLAN and C. ROSSETTI: Lectures at the Symposium on Weak Interactions of the Hungarian Physical Society, Balatonvilagos, 1966.
152. G. FURLAN and C. ROSETTI: *Phys. Lett.* **23**, 499 (1966).
153. M. K. GAILLARD: *Phys. Lett.* **20**, 533 (1966).
154. S. GASIOROWICZ: *Phys. Rev.* **146**, 1067 (1966).
155. S. GASIOROWICZ: *Phys. Rev.* **146**, 1071 (1966).
156. S. GASIOROWICZ and D. A. GEFFEN: *Phys. Lett.* **22**, 344 (1966).
157. S. GASIOROWICZ and D. A. GEFFEN: Minnesote preprint: *A Sum Rule for the $\rho\omega\pi$ coupling Constant.*
158. R. GATTO, L. MAIANI and G. PREPARATA: *Nuovo Cimento* **41A**, 622 (1966).
159. R. GATTO, L. MAIANI and G. PREPARATA: *Phys. Rev. Lett.* **16**, 377 (1966).
160. R. GATTO, L. MAIANI and G. PREPARATA: *Physics* **3**, 1 (1967).
161. R. GATTO, L. MAIANI and G. PREPARATA: *Phys. Lett.* **21**, 459 (1966).
162. R. GATTO, L. MAIANI and G. PREPARATA: *Nuovo Cimento* **44A**, 1279 (1966).
163. D. A. GEFFEN: Minnesota preprint COO-1371-33.
164. M. GELL-MANN: *Phys. Rev.* **125**, 1067 (1962).
165. M. GELL-MANN: *Phys. Lett.* **8**, 214 (1964).
166. M. GELL-MANN: *Physics* **1**, 63 (1964).
167. M. GELL-MANN: *Phys. Rev. Lett.* **14**, 77 (1965).
168. M. GELL-MANN and Y. NE'EMAN: *Ann. Phys.* **30**, 360 (1964).
169. I. S. GERSTEIN: *Phys. Rev. Lett.* **16**, 114 (1966).
170. I. S. GERSTEIN and B. W. LEE: *Phys. Rev.* **144**, 1142 (1966).
171. I. S. GERSTEIN and B. W. LEE: *Phys. Rev. Lett.* **16**, 1060 (1966).
172. L. S. GERSTEIN and B. W. LEE: *Phys. Rev.* **152**, 1418 (1966).
173. J. L. GERVAIS, J. ILIOPOULOS and J. M. KAPLAN: *Phys. Lett.* **21**, 103 (1966).
174. J. L. GERVAIS and M. LeBELLAC: *Nuovo Cimento* **47A**, 822 (1967).
175. F. J. GILMAN and H. J. SCHNITZER: *Phys. Rev.* **150**, 1362 (1966).
176. A. M. GLEESON: *Phys. Rev.* **149**, 1242 (1966).
177. T. GOTO and T. IMAMURA: *Prog. Theor. Phys.* **14**, 396 (1955).
178. (R) M. GOURDIN: Lecture at the Moriond Meeting on Electromagnetic Interactions, 1965/6.
179. M. GOURDIN: *Nuovo Cimento* **47A**, 145 (1967).
180. M. GOURDIN: *Nuovo Cimento* **47A**, 195 (1967).
181. M. GOURDIN: *Phys. Lett.* **22**, 340 (1966).
182. (R) M. GOURDIN: Orsay preprint TH 161.
183. V. N. GRIBOV, B. L. JOFFE and V. M. SHEKHTER: *Phys. Lett.* **21**, 457 (1966).

184. R. W. Griffith: *Phys. Rev.* **147**, 1141 (1966).
185. B. Hamprecht: *Nuovo Cimento* **47A**, 770 (1967).
186. B. Hamprecht: Torino preprint: *Schwinger Terms in Perturbation Theory*.
187. Y. Hara: ICTP Trieste preprint IC/66/44.
188. Y. Hara: *Phys. Rev.* **150**, 1175 (1966).
189. Y. Hara and Y. Nambu: *Phys. Rev. Lett.* **16**, 875 (1966).
190. Y. Hara, Y. Nambu and J. Schechter: *Phys. Rev. Lett.* **16**, 380 (1966).
191. H. Harari: *Phys. Rev. Lett.* **16**, 964 (1966).
192. H. Harari: *Phys. Rev. Lett.* **17**, 56 (1966).
193. H. Harari: Stanford preprint SLAC-58.
194. G. H. Henderson: Edinburgh preprint: *An SU_6 Result from Chiral $SU_2 \times SU_2$*.
195. P. W. Higgs: North Carolina preprint: *The Assignment of Particles to Multiplets of Chiral $SU_3 \times SU_3$*.
196. D. Horn: *Phys. Rev. Lett.* **17**, 778 (1966).
197. M. Hosoda and K. Yamamoto: *Prog. Theor. Phys.* **36**, 425 (1966).
198. M. Hosoda and K. Yamamoto: *Prog. Theor. Phys.* **36**, 426 (1966).
199. S. Humble and F. Poole: *Nuovo Cimento* **45A**, 758 (1966).
200. R. C. Hwa and J. Nuyts: *Phys. Rev.* **151**, 1215 (1966).
201. C. Itzykson and M. Jacob: Saclay preprint: *Non Leptonic Hyperon Decays*.
202. S. Iwao: Bern preprint: *Photonic Decay of Hyperons and the Suzuki-Sugawara Hamiltonian*.
203. S. Iwao: Bern preprint: *Axial-Vector Coupling Constant Renormalization and Weak Interactions*.
204. S. Iwao: Bern preprint: *Consistency of the Approach in Current Algebras*.
205. S. Iwao: Bern preprint: *Soft-Pion Production and Algebra of Currents*.
206. Y. Iwasaki: *Nuovo Cimento* **42A**, 198 (1966)
207. Y. Iwasaki, S. Matsuda and K. Nakamura: *Nuovo Cimento* **43A**, 210 (1966).
208. M. Jacob: Saclay preprint: *Algèbre des courants et singulantés proches*.
209. M. Jacob and G. Mahoux: *Nuovo Cimento* **47A**, 742 (1967).
210. K. Johnson: *Nucl. Phys.* **25**, 431 (1962).
211. K. Johnson and F. E. Low: *Suppl. Prog. Theor. Phys.* **37**, 74 (1966).
212. R. E. Kallosh: *Phys. Lett.* **22**, 519 (1966).
213. A. N. Kamal: *Phys. Rev.* **150**, 1392 (1966).
213a. A. Katz: *Nuovo Cimento* **45A**, 721 (1966).
214. J. C. Katzin: Maryland Report NO. 537.
215. K. Kawarabayashi, W. D. McGlinn and W. W. Wada: *Phys. Rev. Lett.* **15**, 897 (1965).
216. K. Kawarabayashi and M. Suzuki: *Phys. Rev. Lett.* **16**, 255 (1966); E **16**, 384.
217. K. Kawarabayashi and M. Suzuki: *Phys. Rev.* **150**, 1181 (1966).
218. K. Kawarabayashi and M. Suzuki: *Phys. Rev.* **152**, 1383 (1966).
219. K. Kawarabayashi and W. W. Wada: *Phys. Rev.* **137**, B1002 (1965).
220. K. Kawarabayashi and W. W. Wada: *Phys. Rev.* **146**, 1209 (1966).
221. K. Kawarabayashi and W. W. Wada: *Phys. Rev.* **152**, 1286 (1966).
222. A. Katz: *Nuovo Cimento* **45A**, 721 (1966).
223. B. H. Kellett: Imperial College London preprint: ICTP /66/26.
224. C. W. Kim and H. Primakoff: *Phys. Rev.* **147**, 1034 (1966).
225. M. P. Khanna and S. Okubo: *Nuovo Cimento* **44A**, 229 (1966).
226. M. P. Khanna and S. Okubo: *Ann. Phys.* **40**, 153 (1966).

227. M. P. KHANNA and S. OKUBO and N. MUKUNDA: *Nuovo Cimento* **43A**, 33 (1966).
228. M. P. KHANNA, and A. VAIDYA: ICTP Trieste preprint IC/66/47.
229. N. N. KHURI: *Phys. Rev.* **153**, 1477 (1967).
230. K. KIKKAWA: *Prog. Theor. Phys.* **35**, 304 (1966).
231. S. KITAKADO and A. TOYODA: *Prog. Theor. Phys.* **36**, 653 (1966).
232. G. KRAMER and K. MEETZ: DESY preprint 66-8.
233. W. KROLIKOWSKI: ICTP Trieste preprint IC/66/8.
234. W. KROLIKOWSKI: ICTP Trieste preprint IC/66/70.
235. H. KÜHNELT: Graz preprint: 23/1966.
236. W. KUMMER, H. PIETSCHMANN and A. P. BALACHANDRAN: *Ann. Phys.* **29**, 161 (1964).
237. T. K. KUO and M. SUGAWARA: *Phys. Rev.* **151**, 1181 (1966).
238. J. KWIECINSKI: Cracow preprint: PAK 480.
239. J. LANGERHOLC: DESY preprint 66–24.
240. B. W. LEE: *Phys. Rev. Lett.* **14**, 677; E **14**, 850 (1965).
241. C. A. LEVINSON and I. T. MUZINICH: *Phys. Rev. Lett.* **15**, 715; E **15**, 842 (1965).
242. H. T. LIPKIN: Lecture at Yalta International School, Yalta, 1966.
243. H. T. LIPKIN, H. R. RUBINSTEIN and S. MESHKOV: *Phys. Rev.* **148**, 1405 (1966).
244. T. P. LOUBATON and G. MENNESSIER: *Nuovo Cimento* **46A**, 328 (1967).
245. E. Y. C. LU: Cambridge preprint: *Sum Rules and Schwinger Terms in Equal Time Commutators.*
246. E. LU and M. TAHA: *Phys. Rev.* **152**, 1507, (1966).
247. D. H. LYTH: *Phys. Lett.* **21**, 338; E **21**, 719 (1966).
248. K. T. MAHANTHAPPA and RIAZUDDIN: *Nuovo Cimento* **45A**, 252 (1966).
249. H. S. MANI, GYAN MOHAN and L. K. PANDE: *Nuovo Cimento* **44A**, 265 (1966).
250. R. E. MARSHAK, V. S. MATHUR and L. K. PANDIT: *Phys. Lett.* **21**, 563 (1966).
251. V. S. MATHUR, S. OKUBO and L. K. PANDIT: *Phys. Rev. Lett.* **16**, 371 (1966).
252. V. S. MATHUR and L. K. PANDIT: *Phys. Lett.* **19**, 523 (1965).
253. V. S. MATHUR and L. K. PANDIT: *Phys. Rev.* **143**, 1216 (1966).
254. V. S. MATHUR and L. K. PANDIT: *Phys. Lett.* **20**, 308 (1966).
255. V. S. MATHUR and L. K. PANDIT: *Phys. Rev.* **147**, 965 (1966).
256. V. S. MATHUR, L. K. PANDIT and R. E. MARSHAK: *Phys. Rev. Lett.* **16**, 947 (1966), E **16**, 1135.
257. V. S. MATHUR, L. K. PANDIT and RIAZUDDIN: *Phys. Rev. Lett.* **17**, 736 (1966).
258. V. S. MATHUR, L. K. PANDIT and RIAZUDDIN: Rochester preprint: UR 875–170.
259. S. MATSUDA: *Prog. Theor. Phys.* **35**, 858 (1966).
260. S. MATSUDA: *Nuovo Cimento* **44A**, 448 (1966). E **46A**, 814 (1966).
261. S. MATSUDA: *Prog. Theor. Phys.* **36**, 1227 (1966).
262. V. A. MATVEEV, V. G. PISARENKO and B. V. STRUMINSKY: Dubna preprint E-2822.
263. V. A. MATVEEV, B. V. STRUMINSKY and A. N. TAVKHELIDZE: *Phys. Lett.* **23**, 145 (1966).
264. (R) Ph. MEYER: Lectures at the Ecole d'été de physique théorique, Cargèse, 1966.

265. J. W. Moffat: *Phys. Rev.* **145**, 1777 (1966).
266. J. W. Moffat: *Phys. Lett.* **23**, 148 (1966).
267. J. W. Moffat: Toronto preprint: *Asymptotic Theorem Based on Current Algebra and the Quark Model.*
268. (R) J. W. Moffat: Lectures at the: V. Internationale Universitätswochen für Kernphysik, Schladming, 1966.
269. (R) J. W. Moffat: Lectures at the Nato Advanced Study Institute on Symmetry Principles and Fundamental Particles, Istanbul, 1966.
270. J. W. Moffat and P. J. O'Donnel: Toronto preprint: *Symmetry Breaking Corrections to Electromagnetic Sum Rules.*
271. N. Mukunda and T. K. Radha: *Nuovo Cimento* **44A**, 726 (1966).
272. G. Murtaza and P. J. O'Donnel: Toronto preprint: *Renormalization of the Axial Vector Coupling Constant and SU_3 Invariance.*
273. G. Murtaza and P. J. O'Donnel: Toronto preprint: *Non-Leptonic Decays of Hyperons.*
274. I. J. Muzinich: *Phys. Rev.* **151**, 1206 (1966).
275. I. J. Muzinich and S. Nussinov: *Phys. Lett.* **19**, 248 (1965).
276. P. Narayanaswamy, T. Pradhan and T. S. Santhanam: ICTP Trieste preprint IC/66/3.
277. M. Nauenberg: *Phys. Lett.* **22**, 201 (1966).
278. (R) Y. Ne'eman: Lectures at the Pacific Summer School in Physics, Honolulu, 1965.
279. B. M. K. Neffkens: *Phys. Lett.* **22**, 94 (1966).
280. R. Odorioco: Trieste University preprint: *Connections between the Superconvergent Sum Rules and the Current Algebra Approach.*
281. R. Odorioco: Trieste University preprint: *Strong Interaction Sum Rules and Helicity Amplitudes.*
282. R. Oehme: *Ann. Phys.* **33**, 108 (1965).
283. R. Oehme: *Phys. Rev.* **143**, 1138 (1966).
284. R. Oehme: *Nuovo Cimento* **45A**, 666 (1966).
285. R. Oehme: *Phys. Rev.* **148**, 1537 (1966).
286. R. Oehme: *Phys. Lett.* **21**, 567 (1966).
287. R. Oehme: *Phys. Lett.* **22**, 206 (1966).
288. R. Oehme: *Phys. Rev. Lett.* **16**, 215 (1965).
289. R. Oehme: Chicago preprint: EFINS 66–84.
290. R. Oehme and G. Segrè: *Phys. Lett.* **11**, 94 (1964).
291. S. Okubo: *Phys. Lett.* **17**, 172 (1965).
292. S. Okubo: *Phys. Lett.* **20**, 195 (1966).
293. S. Okubo: *Ann. Phys.* **38**, 377 (1966).
294. (R) S. Okubo: *Suppl. Prog. Theor. Phys.* **37**, 114 (1966).
295. S. Okubo: *Nuovo Cimento* **41A**, 586 (1966).
296. S. Okubo: *Nuovo Cimento* **42A**, 1029 (1966).
297. S. Okubo: *Nuovo Cimento* **44A**, 276 (1966).
298. S. Okubo: ICTP Trieste preprint IC/66/10.
299. S. Okubo, R. E. Marshak, and H. Goldberg: Rochester preprint UR–875–120.
300. N. Panchapakesan: Chicago preprint: EFINS 66–35.
301. (R) L. K. Pandit: Lecture at the Third Coral Gables Conference on Symmetry Principles at High Energy, 1966.
302. L. K. Pandit, and I. Schechter: *Phys. Lett.* **19**, 56 (1965).
303. (R) N. J. Papastamatiou: Rutherford Laboratory Report.
304. I. Pasupathy: Nuovo Cimento **45A**, 780 (1966). E **46A**, 814 (1966).

305. I. PASUPATHY and R. E. MARSHAK: *Phys. Rev. Lett.* **17**, 888 (1966).
306. I. PATSAKOS, G. SEGRÈ and J. D. WALECKA: *Phys. Lett.* **23**, 141 (1966).
307. (R) H. PIETSCHMANN: Göteborg preprint: *Lecture Notes on Current Algebras.*
308. K. POHLMEYER: *Nuovo Cimento* **43A**, 762 (1966).
309. K. POLYAKOV: *JETP Lett.* **4**, 50 (1966).
310. T. PRADHAN: *Nucl. Phys.* **9**, 124 (1958).
311. S. RAGUSA: *Nuovo Cimento* **46A**, 514 (1966).
312. R. RAJARAMAN: *Phys. Lett.* **22**, 102 (1966).
313. R. RAJARAMAN: *Phys. Rev.* **145**, 1164 (1966).
314. K. RAMAN: *Phys. Rev. Lett.* **17**, 983 (1966). E **17**, 1238, E **18**, 432.
315. K. RAMAN: Syracuse preprints: NYO–3399–70 and NYO–3399–71.
316. K. RAMAN and E. C. G. SUDARSHAN: *Phys. Lett.* **21**, 450 (1966).
317. K. RAMAN and E. C. G. SUDARSHAN: Syracuse preprint NYO–3399–66.
318. B. RENNER: *Phys. Lett.* **20**, 70 (1966).
319. B. RENNER: *Nucl. Phys.* **87**, 337 (1966).
320. B. RENNER: *Phys. Lett.* **21**, 143 (1966).
321. B. RENNER: *Nuovo Cimento* **43A**, 1154 (1966).
322. B. RENNER: *Nuovo Cimento* **45A**, 689 (1966).
323. B. RENNER: Rutherford Laboratory Report RHEL/R 126.
333. B. RENNER: Cambridge preprint: *Vertex Functions in the Dispersion Theory of Current Algebras.*
334. (Withdrawn.)
335. RIAZUDDIN and FAYYAZUDDIN: *Phys. Rev.* **147**, 1071 (1966).
336. RIAZUDDIN and FAYYAZUDDIN: *Nuovo Cimento* **44A**, 546 (1966).
337. RIAZUDDIN and B. W. LEE: *Phys. Rev.* **146**, 1202 (1966): E **150**, 1406 (1966).
338. RIAZUDDIN and K. T. MAHANTHAPPA: *Phys. Rev.* **147**, 972 (1966).
339. D. W. ROBINSON: Lectures at the Nato Advanced Study Institute on Symmetry Principles and Fundamental Particles, Istanbul, 1966.
340. R. ROCKMORE: *Phys. Rev.* **153**, 1490 (1967).
341. C. RYAN: *Ann. Phys.* **38**, 1 (1966).
342. C. RYAN: *Phys. Rev.* **140B**, 480 (1965).
343. C. RYAN: *Phys. Rev.* **147**, 1139 (1966).
344. (R) C. RYAN and K. WATANABE: Dublin preprint: K_{l3} *Decay Form Factors, Renormalization Effects and the Cabibbo Angle.*
345. J. J. SAKURAI: *Phys. Rev. Lett.* **17**, 552 (1966).
346. J. J. SAKURAI: *Phys. Lett.* **22**, 689 (1966).
347. A. SATO and S. SASAKI: *Prog. Theor. Phys.* **35**, 335 (1966).
348. H. T. SCHNITZER: *Phys. Lett.* **20**, 539 (1966).
349. H. T. SCHNITZER: *Phys. Rev.* **141**, 1484 (1966).
350. B. SCHROER and P. STICHEL: *Comm. Math. Phys.* **3**, 258 (1966).
351. T. SCHWINGER: *Phys. Rev. Lett.* **3**, 296 (1959).
352. T. SCHWINGER: *Phys. Rev.* **130**, 406 (1963).
353. G. SEGRÈ, and T. D. WALECKA: *Ann. Phys.* **40**, 337 (1966).
354. A. SIRLIN: *Phys. Rev. Lett.* **16**, 872 (1966).
355. G. SOLIANI: Torino preprint: *On the Commutation Rules between Axial Charges and Axial Divergences.*
356. L. D. SOLOVIEV: *Sov. Journ. Nucl. Phys.* **3**, 131 (1966).
357. (R) N. STRAUMANN: Lecture at the École d'été de physique théorique, Cargèse, 1965.
358. E. C. G. SUDARSHAN: *Phys. Rev. Lett.* **14**, 1083 (1965).
359. E. C. G. SUDARSHAN: Syracuse preprint NYO–3399–45.

360. H. Sugawara: *Phys. Rev. Lett.* **15**, 870 (1965); E997.
361. P. Suranyi: Budapest preprint: *Electromagnetic Mass Differences and Current Algebras*.
362. D. G. Sutherland: *Nuovo Cimento* **48A**, 188.
363. D. G. Sutherland: *Phys. Lett.* **23**, 384 (1966).
364. H. Suura and L. M. Simmons, Jr.: *Phys. Rev. Lett.* **16**, 598 (1966).
365. H. Suura and L. M. Simmons, Jr.: *Phys. Rev.* **148**, 1579 (1966).
366. M. Suzuki: *Phys. Rev. Lett.* **15**, 986 (1965).
367. M. Suzuki: *Phys. Rev. Lett.* **16**, 212 (1966).
368. M. Suzuki: *Phys. Rev.* **144**, 1154 (1966).
369. J. A. Swieca: *Phys. Rev. Lett.* **17**, 974 (1966).
370. K. Tanaka: *Phys. Rev.* **151**, 1203 (1966).
371. J. C. Taylor: *Phys. Lett.* **22**, 665 (1966).
372. J. C. Taylor: Oxford preprint: *Algebra of Currents, Sum Rules and Vector Mesons*.
373. Y. Tomozawa: *Nuovo Cimento* **46A**, 707 (1966).
374. A. I. Vainshtein, V. V. Sokolov and I. B. Khriplovich: *JETP Lett.* **4**, 132 (1966).
375. N. Veltman: *Phys. Rev. Lett.* **17**, 553 (1966).
376. N. Veltman: Lecture at the Meeting of the Royal Society, London, 2. Nov. 1966.
377. G. Veneziano: *Nuovo Cimento* **43A**, 529 (1966).
378. G. Veneziano: *Nuovo Cimento* **44A**, 295 (1966).
379. W. W. Wada: *Phys. Rev. Lett.* **16**, 956 (1966).
380. K. Watanabe: *Nuovo Cimento* **42A**, 707 (1966).
381. S. Weinberg: *Phys. Rev. Lett.* **16**, 879 (1966).
382. S. Weinberg: *Phys. Rev. Lett.* **17**, 336 (1966).
383. S. Weinberg: *Phys. Rev. Lett.* **17**, 616 (1966).
384. S. Weinberg: *Phys. Rev.* **150**, 1313 (1966).
385. W. I. Weisberger: *Phys. Rev. Lett.* **14**, 1047 (1965).
386. W. I. Weisberger: *Phys. Rev.* **143**, 1302.
387. W. I. Weisberger: *Proc. of the Int. Conf. on Weak Interactions*, Argonne, 1965, p. 409.
388. J. Weyers: Ohio preprint COO–1573–3.
389. J. Weyers, L. L. Foldy and D. R. Speiser: *Phys. Rev. Lett.* **17**, 1062 (1966).

Extra references added in proof

390. H. D. I. Abarbanel: *Phys. Rev.* **153**, 1547 (1967).
391. M. Ademollo, R. Gatto, E. Longhi, G. Veneziano: *Nuovo Cimento* **47A**, 334 (1967).
392. S. L. Adler: *Phys. Rev. Lett.* **18**, 519 (1967).
393. S. L. Adler, Y. Dothan: *Phys. Rev.* **151**, 1267 (1966).
394. V. de Alfaro, S. Fubini, G. Furlan, C. Rossetti: Torino preprint: *Superconvergence and Current Algebra* (1966).
395. G. Altarelli, F. Buccella, R. Gatto: *Phys. Lett.* **24B**, 57 (1967).
396. (R) D. Amati: Lectures at the International School of Elementary Particle Physics, Herzeg Novi, September 1966.
397. D. Amati, R. Jengo: *Phys. Lett.* **24B**, 108 (1967).
398. D. Amati, R. Jengo, E. Remiddi: CERN preprint TH759 (1967).
399. N. Angelescu, E. Rădescu: *Nucl. Phys.* **B1**, 197 (1967).
400. R. D'Auria, V. de Alfaro: *Nuovo Cimento* **48A**, 284 (1967).
401. P. Babu, F. J. Gilman, M. Suzuki: *Phys. Lett.* **24B**, 65 (1967).

402. K. Bardakci, G. Segrè: Berkeley preprint: *On the Algebraic Structure of Superconvergence Relations* (1966).

403. K. J. Barnes, R. Williams, E. Kazes, J. Paton: Imperial College London preprint ICTP/67/8 (1967).

404. T. Becherrawy: Rochester preprint UR–875–174 (1966).

405. C. G. Bollini, J. J. Giambiagi: *Nucl. Phys.* **87**, 465 (1966).

406. C. G. Bollini, J. J. Giambiagi: *Nuovo Cimento* **45A**, 1042 (1966).

407. C. Bouchiat, G. Flamand, J. M. Kaplan: Orsay preprint Th 173 (1966).

408. C. Bouchiat, G. Flamand, Ph. Meyer: Orsay preprint Th 187 (1967).

409. D. G. Boulware: *Phys. Rev.* **151**, 1024 (1966).

410. D. G. Boulware, L. S. Brown: preprint: *Vector Fields and Current Commutators* (1966).

411. D. G. Boulware, S. Deser: *Phys. Lett.* **22**, 99 (1966).

412. D. G. Boulware, S. Deser: *Phys. Rev.* **151**, 1278 (1966).

413. R. A. Brandt: Maryland preprint UMD 643 (1966).

414. J. B. Bronzan, I. S. Gerstein, B. W. Lee, F. E. Low: *Phys. Rev. Lett.* **18**, 32 (1966).

415. J. B. Bronzan, I. S. Gerstein, B. W. Lee, F. E. Low: MIT preprint: *Current Algebra and Non-Regge Behaviour of Weak Amplitudes* II (1967).

416. F. Buccella, G. Veneziano, R. Gatto: *Nuovo Cimento* **42A**, 1019 (1966). E **43A**, 768.

417. G. Calucci: *Nucl. Phys.* **87**, 326 (1966).

418. P. Carruthers, H. W. Huang: Cornell preprint: *Current Algebra Description of Double Pion Photoproduction* (1966).

419. Y. T. Chiu, J. Schechter: *Nuovo Cimento* **46A**, 548 (1966).

420. Y. T. Chiu, J. Schechter, Y. Ueda: Chicago preprint EFINS 66–104 (1966).

421. M. Cini, M. De Maria, B. Taglienti: *Nuovo Cimento* **48A**, 277 (1967).

422. R. A. Coleman, J. W. Moffat: Toronto preprint: *Two-Dimensional Relativistic Quark Model with Exact Solutions* (1967).

423. P. D. Conway: *Phys. Lett.* **24B**, 59 (1967).

424. L. A. Copley, D. Masson: Toronto preprint: *Broken Symmetry Sum Rules and the Algebra of Currents* (1966).

425. G. Costa, C. A. Savoy, A. H. Zimerman: *Nuovo Cimento* **47A**, 319 (1967).

426. T. Das, M. Grynberg, K. Kikkawa: Rochester preprint UR 875–113.

427. K. Dietz: *Phys. Rev.* **152**, 1306 (1966).

428. K. Dietz, J. Kupsch: Bonn preprint: *Covariance of the Retarded Product of Axial-Vector Currents and the Form of Their Equal-Time Commutator* (1967).

429. E. Eberle: *Nuovo Cimento* **46A**, 803 (1966).

430. S. Fenster, N. Panchapakesan: Chicago preprint EFINS 66–90 (1966).

431. S. Fenster, N. Panchapakesan: preprint: *Nonleptonic Hyperon Decay and Baryon-Weak Hamiltonian Commutators* (1966).

432. M. Fitelson, E. Kazes: Pennsylvania preprint: $K_{\mu 3}$ and K_{e3} *Form Factors at Finite Momentum Transfers* (1966).

433. A. Frenkel, M. Posch, G. Suranyi, P. Suranyi: *Nuovo Cimento* **47A**, 626 (1967).

434. N. H. Fuchs: *Phys. Rev. Lett.* **18**, 373 (1967).

435. R. Gatto, L. Maiani, G. Preparata: *Phys. Rev. Lett.* **18**, 97 (1967).

436. D. A. Geffen: MIT preprint: *The Role of Axial Mesons in Current Algebra* (1967).

437. M. GELL-MANN: Lectures at the International School of Physics Ettore Majorana, Erice 1966.

438. F. GHIELMETTI, S. IWAO: Bern preprint: *Three Body Photonic Decay and Algebra of Currents* (1967).

439. R. H. GRAHAM, L. O'RAIFEARTAIGH, S. PAKVASA: Syracuse preprint: NYO–3399–88 (1966).

440. D. F. GREENBERG: Chicago preprint EFINS 66–69 (1966).

441. K. C. GUPTA: *Phys. Rev.* **151**, 1250 (1966).

442. G. S. GURALNIK, V. S. MATHUR, L. K. PANDIT: Rochester preprint UR–875–148 (1966).

443. I. G. HALLIDAY: Cambridge preprint: *Sum Rules and Perturbation Theory* (1967).

444. I. G. HALLIDAY, P. V. LANDSHOFF: Cambridge preprint: *Fubini Sum Rules for Vertex Functions* (1967).

445. J. HAMILTON: Copenhagen preprint: *On Zero Mass Pions* (1967).

446. B. HAMPRECHT: Torino preprint: *Schwinger Terms in Perturbation Theory* (1966).

447. Y. HARA: Tokyo (Univ. of Education) preprint: TUEP –66–17 (1966).

448. H. HARARI: *Phys. Rev. Lett.* **17**, 1303 (1966).

449. H. HARARI: *Phys. Rev. Lett.* **18**, 319 (1967).

450. G. HÖHLER, R. STRAUSS: Karlsruhe preprint: *A New Evaluation of the Adler-Weisberger Sum Rule* (1967).

451. G. HÖHLER, R. STRAUSS: Karlsruhe preprint: *Test of a Sum Rule for the Nucleon Magnetic Moments* (1967).

452. (R) C. ITZYKSON, M. JACOB, G. MAHOUX: Saclay preprint: *Non-Leptonic K-Meson Decays* (1966).

453. S. IWAO: *Nucl. Phys.* **B1**, 117 (1967).

454. L. JENKOVSZKY, V. V. KUKHTIN, I. MONTVAY, NGUYEN VAN HIEU: Dubna preprint E2–3039 (1966).

455. H. F. JONES, M. D. SCADRON: Imperial College London preprint: ICTP/66/25 (1966).

456. D. L. KATYAL: *Nuovo Cimento* **47A**, 869 (1967).

457. I. KHAN: Edinburgh preprint: *Superconvergence Relations for Pion Photoproduction and Pion Scattering* (1967).

458. I. KHAN, M. S. K. RAZMI: Edinburgh preprint: *On the Isobar Model of Pion Photoproduction of Hyperons* (1967).

459. C. W. KIM: *Phys. Rev.* **151**, 1216 (1966).

460. C. W. KIM, H. PRIMAKOFF: *Phys. Rev.* **147**, 1034 (1966).

461. C. W. KIM, M. RAM: *Phys. Rev. Lett.* **18**, 327 (1967).

462. J. KWIECINSKI: Cracow preprint PAK–NP 496 PL.

463. M. LÉVY: Paris preprint: *Currents and Symmetry Breaking* (1967).

464. G. LONGHI: *Nuovo Cimento* **47A**, 282 (1967).

465. E. Y. C. LU: Cambridge preprint: *A Theorem on C-Number Schwinger Terms* (1967).

466. G. MACK: Bern preprint: *Partially Conserved Dilation Current* (1967).

467. L. MAIANI, G. PREPARATA: Roma preprint: *A Soft Photon Emission Approach to $\pi^\circ \to 2\gamma$ and $\eta \to \gamma\gamma\pi^\circ$ Decays* (1966).

468. S. MATSUDA: Tokyo preprint: *Superconvergence Condition and Okubo's Ansatz* (1967).

469. S. MATSUDA, S. ONEDA, J. SUCHER: Maryland preprint UMD 633 (1966).

470. J. W. MEYER: *Phys. Rev.* **153**, 1652 (1967).

471. MING CHIANG LI: Princeton Inst. of Adv. Stud. preprint: *Current Algebra and the Tadpole Model in Nonleptonic Decays* (1966).
472. MING CHIANG LI: Princeton Inst. of Adv. Stud. preprint: *Current Algebra and the Tadpole Model in $B' \to B\gamma$ Decays* (1966).
473. A. MINGUZZI, G. VELO: Bologna preprint: *Current Algebra and Unphysical Range in Dispersion Relations* (1966).
474. A. K. MOHANTI: *Nuovo Cimento* **46A**, 556 (1966).
475. V. F. MÜLLER, J. ROTHLEITNER: Heidelberg preprint: *The Adler–Weisberger Sum Rule* (1966).
476. I. J. MUZINICH: *Phys. Rev. Lett.* **18**, 381 (1967).
477. T. NAGYLAKI: Caltech preprint CALT 68–111 (1967).
478. M. NAUENBERG: Stanford preprint SLAC–PUB–212 (1966).
479. M. NAUENBERG: Lectures at the VI. Internationale Universitäts-wochen für Kernphysik, Schladming, 1967.
480. R. OEHME, G. VENTURI: Chicago preprint EFINS 67–4 (1967).
481. P. OLESEN: Copenhagen preprint: *Non-Conservation of the Isovector Current and Electromagnetic Mass Differences* (1967).
482. S. ONEDA, J. C. PATI: Maryland preprint UMD 575 (1966).
483. L. K. PANDE: Trieste preprint: ICTP/67/6 (1967).
484. V. G. PISARENKO: Dubna preprint P2931 (1966).
485. G. POCSIK: *Nuovo Cimento* **43A**, 541 (1966).
486. J. C. POLKINGHORNE: Cambridge preprint: *Schwinger Terms and the Johnson Low Model* (1967).
487. H. PRIMAKOFF: Orsay preprint: Th 171 (1967).
488. (R) L. A. RADICATI: Lectures at the Scottish Universities' Summer School, Edinburgh 1966.
489. S. RAI CHOUDHURY, G. C. JOSHI: *Nuovo Cimento* **47A**, 680 (1967).
490. R. RAMACHANDRAN: *Nuovo Cimento* **47A**, 669 (1967).
491. RIAZUDDIN, FAYYAZUDDIN: Chicago preprint EFINS 66–117 (1966).
492. H. R. RUBINSTEIN, G. VENEZIANO: *Phys. Rev. Lett.* **18**, 411 (1967).
493. H. R. RUBINSTEIN, G. VENEZIANO: Weizmann Inst. preprint: *Connection between Regge Pole Parameters and Local Commutation Relations* (1967).
494. B. SAKITA, K. C. WALI: *Phys. Rev. Lett.* **18**, 29 (1966).
495. J. J. SAKURAI: Chicago preprint EFINS 66–103 (1966).
496. A. SATO, Y. YOKOO, J. TAKAHASHI: Osaka preprint: *Current Algebra and Form Factors of the Axial Vector Current* (1966).
497. D. SCHILDKNECHT: DESY preprint 67–2 (1967).
498. V. SINGH: *Phys. Rev. Lett.* **18**, 36 (1967), E **18**, 300.
499. P. P. SRIVASTAVA: Syracuse preprint NYO 3399–65 (1966).
500. D. SUTHERLAND: CERN preprint TH 761 (1967).
501. M. O. TAHA: Cambridge preprint: *An Approach to Currents and the Derivation of Current Algebra* (1966).
502. M. O. TAHA: Cambridge preprint: *A Method and Some Results in Current Identities* (1967).
503. M. O. TAHA: Cambridge preprint: *Sum Rules and High Energy Behaviour* (1967).
504. K. TANAKA: Ohio preprint COO–1543–13.
505. M. TONIN: *Nuovo Cimento* **47A**, 919 (1967).
506. T. L. TRUEMAN: *Phys. Rev. Lett.* **17**, 1198 (1966).
507. K. V. VASAVADA: Univ. of Connecticut preprint: *Incompatibility of a Superconvergence Relation with Conservation of A-Parity* (1967).

508. S. Weinberg: *Phys. Rev. Lett.* **18**, 188 (1967).
509. S. Weinberg: *Phys. Rev. Lett.* **18**, 507 (1967).
510. P. Winterlitz, A. A. Makarov, V. A. Matvejev, Nguyen van Hieu: *Sov. Journ. Nucl. Phys.* **3**, 672 (1966).
511. E. del Giudice, E. Galzenati: Napoli preprint: *Dispersive Sum Rules and Low Energy Limits* (1967).
512. C. S. Lai: Indiana University preprint: *Algebra of Currents and the Vector-Meson Baryon Coupling Constants in Broken SU_3* (1966).
513. J. W. Moffat: Toronto preprint: *Theory of Symmetry Violation and Gravitation* (1967).
514. J. Pestiau: Louvain preprint: *Radiative Decays of Hadrons and Current Conservation* (1967).
515. S. Okubo: Rochester preprint: UR 875–171 (1966).
516. S. Okubo: Lecture at the Fourth Coral Gables Conference on Symmetry Principles at High Energy (1967).

Selected General References

S 1. Scottish Universities' Summer School 1960: *Dispersion Relations*; edited by G. R. SCREATON; Edinburgh, Oliver & Boyd (1961).

S 2. Scottish Universities' Summer School 1963: *Strong Interactions and High Energy Physics*; edited by R. G. MOORHOUSE; Edinburgh, Oliver & Boyd (1964).

S 3. Varenna Summer School 1963: *Dispersion Relations and their Connexion with Causality*; New York, Academic Press (1964).

S 4. G. F. CHEW: *S-Matrix Theory of Strong Interactions*; New York, W. A. Benjamin Inc. (1962).

S 5. M. JACOB and G. F. CHEW: *Strong-Interaction Physics*; New York, W. A. Benjamin Inc. (1964).

S 6. R. J. EDEN, P. V. LANDSHOFF, D. I. OLIVE and J. C. POLKINGHORNE: *The Analytic S-Matrix*; Cambridge University Press (1966).

S 7. R. H. DALITZ: Les Houches Summer School 1965; Gordon & Breach Publishers, New York, 1966: Review on quark models.

S 8. See, for instance, V. GUPTA, *Phys. Rev. Lett.* **14**, 838 (1965).

S 9. M. GELL-MANN and NE'EMAN: *The Eightfold Way*; New York, W. A. Benjamin Inc. (1964).

S 10. S. S. GERSTEIN and J. B. ZELDOVITCH: *Zh. Eksper. Teor. Fiz.* **29**, 698 (1955).

S 11. R. P. FEYNMAN and M. GELL-MANN: *Phys. Rev.* **109**, 193 (1958).

S 12. N. CABIBBO: *Phys. Rev. Lett.* **10**, 531 (1963).

S 13. See J. C. TAYLOR: *Reports on Progress in Physics* **27**, 407 (1964); and Y. K. LEE, L. W. MO and C. S. WU: *Phys. Rev. Lett.* **10**, 253 (1963).

S 14. N. BRENE, L. VEJE, M. ROOS and C. CRONSTRÖM: *Phys. Rev.* **149**, 1288 (1966).

S 15. M. ADEMOLLO and R. GATTO: *Phys. Rev. Lett.* **13**, 264 (1964).

S 16. J. LEITNER: Syracuse preprint: *An Experimental Review of SU₃*.

S 17. F. GÜRSEY and L. A. RADICATI: *Phys. Rev. Lett.* **13**, 173 (1964).

S 18. A. W. HENDRY: Rutherford Laboratory Report RHEL/R 102.

S 19. H. RUEGG, W. RÜHL and T. S. SANTHANAM: *Helv. Phys. Acta*, to be published.

S 20. C. LOVELACE, R. M. HEINZ and A. DONNACHIE: *Phys. Lett.* **22**, 332 (1966)

S 21. S. FUBINI, Y. NAMBU and V. WATAGHIN: *Phys. Rev.* **111**, 329 (1958).

S 22. G. CHEW, M. GOLDBERGER, F. LOW and Y. NAMBU: *Phys. Rev.* **106**, 1345 (1957).

S 23. B. R. MARTIN: *Nucl. Phys.* **87**, 177 (1966).

S 24. R. HAAG: *Phys. Rev.* **112**, 669 (1958).

K. NISHIJIMA: *Phys. Rev.* **111**, 995 (1958).

W. ZIMMERMANN: *Nuovo Cimento* **10**, 597 (1958).

H. J. BORCHERS: *Nuovo Cimento* **15**, 784 (1960).

S. KAMEFUCHI, L. O'RAIFEARTAIGH and A. SALAM: *Nucl. Phys.* **28**, 529 (1961).

Bibliography

S 25. N. N. Kroll and M. A. Ruderman: *Phys. Rev.* **93**, 233 (1954).
S 26. P. Federbush and K. Johnson: *Phys. Rev.* **120**, 1926 (1960).
S 27. M. Froissart, *Phys. Rev.* **123**, 1053 (1961).
S 28. For an introduction to Regge poles see S. C. Frautschi: *Regge Poles and S-Matrix Theory*; New York, W. A. Benjamin Inc.; for a recent review E. Leader: *Rev. Mod. Phys.* **38**, 476 (1966).
S 29. F. E. Low: *Phys. Rev.* **96**, 1428 (1954).
S 30. M. Gourdin and Ph. Salin: *Nuovo Cimento* **27**, 193 (1966).
S 31. J. S. Bell, J. Steinberger: Oxford International Conference on Elementary Particles, 1965.

Index

176